U0304821

暨南大学本科教材资助项目

Experimental Food Biotechnology

食品生物技术实验教程

彭喜春 张 宁◎主编

暨南大學出版社
JINAN UNIVERSITY PRESS
中国·广州

图书在版编目（CIP）数据

食品生物技术实验教程/彭喜春，张宁主编．—广州：暨南大学出版社，2021.11
ISBN 978-7-5668-3215-3

Ⅰ.①食… Ⅱ.①彭… ②张… Ⅲ.①生物技术—应用—食品工业—实验—教材
Ⅳ.①TS201.2-33

中国版本图书馆CIP数据核字（2021）第170108号

食品生物技术实验教程
SHIPIN SHENGWU JISHU SHIYAN JIAOCHENG
主　编：彭喜春　张　宁

出 版 人：张晋升
责任编辑：曾鑫华　江楷煜　彭琳惠
责任校对：苏　洁　黄亦秋
责任印制：周一丹　郑玉婷

出版发行：暨南大学出版社（510630）
电　　话：总编室（8620）85221601
　　　　　营销部（8620）85225284　85228291　85228292　85226712
传　　真：（8620）85221583（办公室）　85223774（营销部）
网　　址：http：//www.jnupress.com
排　　版：广州市天河星辰文化发展部照排中心
印　　刷：广州市穗彩印务有限公司
开　　本：787mm×1092mm　1/16
印　　张：15.25
字　　数：333千
版　　次：2021年11月第1版
印　　次：2021年11月第1次
定　　价：48.00元

前　言

食品生物技术（food biotechnology）是生物技术在食品原料生产、食品加工和制造中应用的一门学科。它包括食品发酵和酿造等最古老的生物技术加工过程，也包括应用现代生物技术来改良食品原料的加工品质的基因、生产高质量的农产品、制造食品添加剂、培养植物和动物细胞以及改良与食品加工和制造相关的其他生物技术，如酶工程和蛋白质工程等。生物技术在食品工业中的应用日益广泛和深入，近年来由于组学技术的发展，生物技术与食品工业的融合得到了突飞猛进的发展。“食品生物技术”课程在本科教学中也受到了广大师生的青睐，但在食品科学与工程专业的教学中，并没有相关的实验教程能让学生体验、学习和熟练掌握生物技术在食品工业中的系列应用技术。为弥补这一空白，编者结合十余年的一线教学经验编写了本实验教程，以满足本科甚至研究生的教学。

传统的食品科学与工程专业的课程体系中，很少有“分子生物学”“基因工程”和“细胞生物学”的基础课程，因此也缺乏相关生物技术实验技能的学习。随着生物技术在食品工业中渗透和融合的深入，这种传统的教学体系已经不能满足社会对生物技术实验技能的需求。本书以《食品生物技术导论》的教学内容为基础，组织了与食品工业有关的基因工程、酶工程、蛋白质工程、发酵工程以及细胞工程五个模块的实验内容。该五个模块中，基因工程和细胞工程两部分提供的是基本实验技能的学习；酶工程、蛋白质工程以及发酵工程编入更高层次的实验内容。本书旨在为食品科学与工程学科领域的学生弥补基因工程和细胞工程等基本实验技能的基础，同时为学生提供探究生物技术在食品工业应用中更为深入的酶工程、蛋白质工程和发酵工程技术的方向。本书呈现的实验内容有浅有深、深浅结合，可以为不同的教学目的和要求提供指导。

感谢参与本书编写工作的韩露露、胡庆、孙晓燕、周俊华、柳杨含羞、张晓梅、程婧、余荣轩和李天行同学。尽管我们力求把本书做得更好，但因编者水平有限，尤其是蛋白质工程领域知识的不足，不当之处实在难免，望读者给予批评、指正。

编　者

2021 年 4 月

Preface

Food biotechnology is a subject of biotechnology application in the food raw materials production, processing and manufacturing of food. It includes the oldest biotechnology such as food fermentation and brewing, as well as the application of modern biotechnology to improve the processing quality of food raw materials, the production of high-quality agricultural products, the manufacture of food additives, the cultivation of plants and animal cells, and other biotechnology related to food processing and manufacturing, such as enzyme engineering and protein engineering. Biotechnology is widely used in food industry. In recent years, due to the development of omics technology, the integration of biotechnology and food industry has been developed rapidly. The course of "Food Biotechnology" has also been favored by teachers and students in undergraduate teaching. However, in the traditional teaching plan of food science and engineering, there is no relevant experimental course to let students experience, learn and master the series of biotechnology application skills in food industry. In order to make up for this gap, the author edited this experimental book with more than ten years of teaching experience in order to meet the needs of undergraduate and even graduate students.

In the traditional curriculum system of food science and engineering, there are few basic courses of "Molecular Biology" "Genetic Engineering" and "Cell Biology", so there is also a lack of relevant biotechnology experimental skills. With the in-depth penetration and integration of biotechnology in the food industry, this traditional teaching system is losing its capacity to meet the needs of society for biotechnology experimental skills. Based on the teaching content of *Introduction to Food Biotechnology*, this book organizes the experimental contents of five modules related to food industry, including genetic engineering, enzyme engineering, protein engineering, fermentation engineering and cell engineering. In the five modules, genetic engineering and cell engineering present basic experimental skills; enzyme engineering, protein engineering and fermentation engineering introduce higher level experimental skills. This book aims to provide students in the field of food science and engineering with basic experimental skills such as genetic engineering and cell engineering, and provide them with more in-depth enzyme engineering, protein engineering and fermentation engineering techniques in the application of biotechnology in food industry. The experimental contents presented in this book integrate easy and difficult knowledge, which can provide guidance for different teaching purposes and requirements.

Thanks to Han Lulu, Hu Qing, Sun Xiaoyan, Zhou Junhua, Liu Yanghanxiu, Zhang Xiaomei, Cheng Jing, Yu Rongxuan and Li Tianxing who participated in the compilation of this book. Although we subjectively strive to make this book more comprehensive and complete, due to the limited level of the editor, especially the lack of knowledge in the field of protein engineering, it is inevitable that there are some improper parts in the book. we hope the readers will criticize and correct the improper parts in the book.

Editors

April, 2021

目　录
Contents

一、大肠杆菌基因组 DNA 的提取及其纯度测定

【实验目的】

（1）掌握移液枪的使用方法。

（2）掌握基因组 DNA 的提取和提纯方法。

（3）掌握利用紫外光谱测定 DNA 的浓度和纯度的方法。

【实验原理】

用热裂解法裂解细菌细胞，让基因、蛋白质和其他物质释放出来，冰浴后通过离心沉淀细胞壁和一些蛋白，基因组将悬浮于上清液中。采用这种方法提纯时，上清液中还存在一些蛋白，可能会影响到聚合酶链式反应（PCR）的结果。该方法只适用于革兰氏阴性细菌。

DNA 样品纯度的测定原理是基于 DNA（或 RNA）分子中碱基的紫外吸收图谱在 260 nm处有一个特征吸收峰。这种吸收和溶液中的 DNA（或 RNA）量成正比。在 1 cm 光路的石英比色皿中，DNA 在 260 nm 波长处的消光系数为 20。基于此消光系数，50 μg/mL 双链 DNA 溶液在 260 nm 波长处吸收值为 1。

DNA 的双螺旋和单链分子之间的相互转换会使其吸收水平产生一定的变化。但是这种偏差可以用一个特定的公式校正。这种方法方便且相对准确。

蛋白质对光的最大吸收大约在 280 nm 波长处，这主要归因于色氨酸残基对光的吸收。因此，A_{260}/A_{280}的比值可用来衡量 DNA 的纯度，并且比值应在 1.65 ~ 1.85 之间。如果比值较低，说明 DNA 被蛋白质污染。由于苯酚在 270 nm 波长处有最大吸收，如果 DNA 样品被苯酚污染，则 A_{260}将会异常高，会导致 DNA 浓度被高估。

【实验仪器、材料与试剂】

1. 仪器

低温离心机、恒温水浴箱、台式离心机、移液枪（10 μL、100 μL、1 000 μL 各一支）、紫外分光光度计、彼此配对的 1 cm 石英比色皿、旋涡混合器。

2. 材料

已培养的大肠杆菌菌液、三羟甲基氨基甲烷（Tris）、乙二胺四乙酸（EDTA）、氯化

钠（NaCl）、无水乙醇、十二烷基磺酸钠（SDS）、50 mL 离心管、吸头、小试管、1.5 mL 微量离心管（EP 管）。

3. 试剂

TE 缓冲液：10 mmol/L Tris · HCl （pH 8.0），1 mmol/L EDTA （pH 8.0），1.25% SDS，75% 乙醇。

【实验步骤】

（1）用移液枪取 1.2 mL 培养物于 1.5 mL EP 管中，以 14 000 转/分钟转速离心 5 分钟。用移液枪移除上清液。

（2）取 500 μL TE 缓冲液于上述离心管中，用旋涡混合器震荡使沉淀悬浮均匀。然后 1 400 转/分钟离心 2 分钟，用移液枪移除上清液。

（3）加入 200 μL TE 缓冲液，盖好离心管管盖，用旋涡混合器震荡成悬浮液，置于 100 ℃水浴箱中煮沸 15 分钟，取出后立即放置于冰上冷却 5 ~ 10 分钟，然后 14 000 转/分钟离心 5 分钟（冷却时可置于冰箱下层冷冻间）。

（4）取上清液 150 μL 于新 EP 管中，此即为该细菌的基因组 DNA。

（5）按 1 : 20 或者更高的比例将样品 DNA 溶液在 TE（或 ddH_2O）中稀释。用 TE（或 ddH_2O）在紫外分光光度计 260 nm、280 nm 和 310 nm 波长处做空白校正。

（6）将各 DNA 稀释溶液装入比色皿比色，读取上述三个波长下的光密度（OD）。

（7）记录光密度数据，通过计算确定 DNA 的浓度和纯度。

对于单链 DNA 分子，其浓度（μg/mL）可以通过以下公式导出：

【ssDNA】 $=33\ (OD_{260}-OD_{310})\ \times$稀释倍数

对于双链 DNA 分子，其浓度（μg/mL）可以通过以下公式导出：

【dsDNA】 $=50\ (OD_{260}-OD_{310})\ \times$稀释倍数

对于单链 RNA 分子，其浓度（μg/mL）可以通过以下公式导出：

【ssRNA】 $=40\ (OD_{260}-OD_{310})\ \times$稀释倍数

【注意事项】

（1）使用移液枪时，要看清量程，选择合适的移液器和枪头。吸取液体时，速度要适宜，过快的吸取速度容易使样品进入套柄，损伤仪器，同时也造成样品的交叉污染。

（2）UV 分光光度法可以精确快速并且无破坏性地测定 DNA 的浓度，最低浓度可达 2.5 μg/mL，操作一定要专业精准。提前预热分光光度计，清洗干净比色皿再装入待测液。同时，由于玻璃和塑料会吸收紫外光，本实验必须使用石英比色皿。

（3）OD_{310}的值为背景。如果盐的浓度很高，它的值也会很高。

（4）DNA的OD_{260}：OD_{280}比值应该在1.8左右。高于或低于该值的比率通常表示DNA样本中存在RNA或蛋白质污染。

（5）RNA的OD_{260}：OD_{280}比值约为2.0。

【实验结果】

波长（nm）	T%	光密度（OD）
260		
280		
310		

Ⅰ. Extraction and Purity Determination of *E. coli* Chromosomal DNA

【Objectives】

(1) Learn the method of utilizing a pipette.

(2) Learn the method of isolating and purifying chromosomal DNA from bacterial cells.

(3) Learn the method of determining the concentration and purity of DNA by UV spectroscopy.

【Principle】

In the procedure described here, the bacterial cells are cracked with hot water. Their chromosomal DNA, proteins and the other components will be released from the cells at the same time. After ice bath and centrifugation, the fraction of cell wall will precipitate, and DNA will dissolve and suspend in the water along with part of proteins. PCR may be effected because of the proteins dissolved in the DNA solution. This chromosomal DNA obtained from this method is only suitable for gram-negative bacteria.

The bases in DNA (or RNA) molecules have a characteristic UV-absorption pattern with a peak at 260 nm. This absorption is proportional to the amount of DNA (or RNA) in the solution. The extinction coefficient is 20 of DNA at 260 nm wavelengths in the 1 cm light path of quartz cuvette. Based on this extinction coefficient, the absorption value is 1 for 50 μg/mL dsDNA solution at 260 nm wavelengths.

The mutual transformation between double helix and single stranded molecule of DNA will change its absorption level. But this deviation can be corrected by a specific formula. This method is convenient and relatively accurate.

The maximum absorption of light by protein is at about 280 nm wavelengths, mainly due to the absorption of light by tryptophan residues. The ratio of A_{260}/A_{280}, therefore, can be used to measure the purity of DNA, and the ratio should fall between 1.65 and 1.85. A low ratio refers that the DNA is contaminated by protein. The absorption of phenol can reach its peak at 270 nm wavelengths, therefore, if the DNA sample is polluted by phenol, A_{260} will be abnormally high, which will lead to overestimation of DNA concentration.

【Apparatus, materials and reagents】

1. Apparatus

Refrigerated centrifuge, thermostat water bath cauldron, table centrifuge, three pipettes with the capacities of 10 μL , 100 μL and 1 000 μL respectively, UV spectrophotometer (UV lamp pre-warmed), coupled 1 cm quartz cuvettes, vortex mixer.

2. Materials

The cultured bacterial solution of *E. coli*, trihydroxymethyl aminomethane (Tris), ethylenediamine tetraacetic acid (EDTA), sodium chloride (NaCl), anhydrous ethanol, dodecyl sodium sulfate (SDS), 50 mL centrifuging tubes, pipette tips, mini-tubes and 1.5 mL Eppendorf tubes (EP tubes).

3. Reagents

TE buffer: 10 mmol/L Tris · HCl (pH 8.0), 1 mmol/L EDTA (pH 8.0), 1.25% SDS, 75% ethanol.

【Procedures】

(1) Pipette 1.2 mL bacterial culture into EP tubes, centrifuge at 14 000 rpm rotation speed for 5 minutes. Discard the supernatant by using the pipette.

(2) Pipette 500 μL TE buffer into the tubes, shake the sediment into suspension evenly by vortex mixer. Centrifuge in the microcentrifuge at 14 000 rpm rotation speed for 2 minutes. Discard the supernatant by using pipette.

(3) Add 200 μL TE buffer into the tube and cover it. Shake the content into suspension by vortex mixer. Boil it in 100 ℃ water for 15 mins, take it out and cool it down by ice for 5 – 10 minutes, then recentrifuge at 14 000 rpm rotation speed for 5 minutes (After cooled, it can be frozen and stored at refrigerator).

(4) Pipette 150 μL supernatants into a new 1.5 mL EP tube, which is the genomic DNA of the bacteria.

(5) Prepare 1 : 20 dilution of the stock DNA solution in TE buffer (or ddH_2O). Use TE buffer (or ddH_2O) for blank correction at the wavelengths of 260 nm, 280 nm and 310 nm in UV spectrophotometer.

(6) Load each dilution of DNA solution into the cuvettes and read the optical density (OD) at three wavelengths mentioned above.

(7) Record the OD readings, then determine the purity and concentration of DNA by calculation.

For ssDNA, its concentration in microgram per mL can be derived by the formula:

【ssDNA】 $=33\times(OD_{260}-OD_{310})\times$ dilution factor

For dsDNA, its concentration in microgram per mL can be derived by the formula:

【dsDNA】 $=50\times(OD_{260}-OD_{310})\times$ dilution factor

For ssRNA, its concentration in microgram per mL can be derived by the formula:

【ssRNA】 $=40\times(OD_{260}-OD_{310})\times$ dilution factor

【Notes】

(1) Please select a right pipette and tip to transfer solution. When pipetting liquid, the speed should be appropriate. Pipetting too fast may make the sample enter the handle, damage the instrument, and increase the risk of cross contamination.

(2) UV spectrophotometry can determine the concentration of DNA accurately, rapidly and non-destructively, and the minimum concentration can reach 2.5 μg/mL. The operation must be professional and accurate. Preheat the spectrophotometer in advance. Clean the cuvette and then load the solution to be tested. Quartz cuvette must be used here, as the glass or plastic ones will absorb ultraviolet light.

(3) Take value of OD_{310} as the background. It will be high if the salt concentration is high.

(4) The ratio of $OD_{260}:OD_{280}$ for DNA should be around 1.8. The ratio above or below the value usually indicates the presence of RNA or protein contamination in DNA sample.

(5) The ratio of $OD_{260}:OD_{280}$ for RNA is around 2.0.

【Results】

Wavelengths (nm)	T%	Optical Density (OD)
260		
280		
310		

二、碱去垢剂法分离质粒 DNA 小量制备法

【实验目的】

掌握碱去垢剂法分离质粒 DNA 的原理和方法。

【实验原理】

本方法是依据共价闭合环状质粒 DNA 与染色体 DNA 在变性和复性特性之间存在差异进行的。当细胞悬浮于氢氧化钠（NaOH）和十二烷基磺酸钠（SDS）溶液中时，在高 pH（碱）的作用下细胞发生裂解，蛋白质和染色体 DNA 发生变性。加入中和溶液（酸性乙酸钾）并离心后，变性蛋白和染色体 DNA 就会与细胞碎片一起沉淀下来，而质粒 DNA 则留在上清液中。其实，加入碱溶液时，质粒 DNA 也发生变性，但其两条链仍然靠得很近，就像一条链上的两个环链。当加入酸性溶液进行中和时，质粒 DNA 的两条链就分别与其互补链重新退火，进而形成原始状态的质粒。

本实验采用的是一种从少量培养液或大量细胞克隆中分离质粒 DNA 的简便快速方法。这种方法提取的质粒 DNA 纯度很高，无须进一步纯化，就可以用于限制性内切酶酶切、聚合酶链式反应、序列测定等。

【实验仪器、材料与试剂】

1. 仪器

恒温培养箱、恒温摇床、高压灭菌锅、台式离心机、旋涡混合器、移液枪（10 μL、100 μL、1 000 μL 各一支）。

2. 材料

葡萄糖、三羟甲基氨基甲烷（Tris）、乙二胺四乙酸（EDTA）、氢氧化钠（NaOH）、十二烷基磺酸钠（SDS）、乙酸钾、冰乙酸、氯仿、无水乙醇、胰 RNA 酶、氨苄西林、蔗糖、溴酚蓝、苯酚、8－羟基喹啉、β－巯基乙醇、盐酸（HCl）、含 pUC 19 质粒的大肠杆菌、*Eco*R Ⅰ酶、异丙醇、50 mL 离心管、吸头、小试管。

3. 试剂

（1）溶液Ⅰ（须新鲜配制，冰上保存，临用时加热至室温）：50 mmol/L 葡萄糖，

25 mmol/L Tris · HCl（pH 8.0），10 mmol/L 乙二胺四乙酸（EDTA）（pH 8.0）。

（2）溶液Ⅱ：0.4 mol/L NaOH，2% SDS，用前等体积混合。

（3）溶液Ⅲ：配制 60 mL 5 mol/L 乙酸钾，加 11.5 mL 冰乙酸和 28.5 mL 去离子双蒸水，冰上保存。

（4）TE 缓冲液：配制 10 mmol/L Tris · HCl（pH 8.0），加入 1 mmol/L EDTA（pH 8.0）。

（5）70% 乙醇（放 -20 ℃冰箱中，用后即放回）。

（6）胰 RNA 酶溶液：将 RNA 酶溶于 10 mmol/L Tris · HCl（pH 7.5）/15 mmol/L NaCl 中，配成 10 mg/mL 的浓度，于 100 ℃加热 15 分钟，缓慢冷却至室温，保存于 -20 ℃。

（7）凝胶上样缓冲液（6 ×）：配制 40% 蔗糖溶液，添加 0.25% 溴酚蓝。

（8）用 TE 缓冲液饱和的苯酚与氯仿（1∶1）。

【实验步骤】

（1）从 LB 平板上挑取一单菌落，接种于 100 mL 含相应抗生素的 LB 培养液中，37 ℃振荡（约 225 转/分钟）培养过夜。

（2）在微型离心管中加入 1.2 mL 的过夜培养液，12 000 转/分钟离心 30 秒。

（3）移去上清，将沉淀物用 100 μL 含有 4 mg/mL 溶菌酶的溶液Ⅰ重新悬浮（室温下）。在旋涡混合器上将细胞振荡混匀，并确认无粘连的细胞块。

（4）在室温下放置数秒，可见到溶菌酶对细胞的作用：悬浮的细胞变成轻微的乳状，振荡时易附着于试管壁。将试管置于冰浴冷却 1 分钟。

（5）室温下加入 200 μL 溶液Ⅱ，轻微振荡试管使其混合，并置试管于冰浴再冷却 1 分钟。

（6）加入 150 μL 冰冷的溶液Ⅲ，混合，再冰浴 3 分钟，这时会有白色絮状沉淀物形成。

（7）微型离心机 12 000 转/分钟离心 2 分钟。

（8）将上清液移至新的微量离心管中，注意避免带入白色沉淀物。

（9）加入等体积的苯酚与氯仿（1∶1）的混合物。旋涡振荡 1 分钟，室温下 12 000 转/分钟离心 2 分钟。

（10）将上层水相移入新的离心管中，加入等体积氯仿溶液，旋涡混合振荡 30 秒，12 000 转/分钟离心 2 分钟。

（11）将上层水相移入新的离心管中，加入 2 倍体积的冰冷的无水乙醇，混合后于 -20 ℃沉淀 5 分钟。

（12）12 000 转/分钟离心 4 分钟或以上，弃去上清，用冰冷的 70% 的乙醇淋洗沉淀，然后对沉淀进行真空干燥。

（13）将沉淀重新悬浮在 50 μL 含有 50 μg/mL RNA 酶的 TE 缓冲液中。37 ℃保温 5 分钟（如果不用 RNA 酶进行处理，小于 600 bp 的 DNA 片段在琼脂糖凝胶电泳时会被 RNA 所掩盖）。

（14）将小量制备的质粒试管标上日期和内容，－20 ℃保存备用。

【实验结果】

通过本实验得到了小量质粒，将其放入微型试管中并标上日期和内容，放入－20 ℃冰箱保存备用。

Ⅱ. Isolation of Plasmid DNA by the Alkaline-Detergent Method: A Miniprep Procedure

【Objective】

Master the principle and method of separating plasmid DNA by alkaline-detergent method.

【Principle】

This method exploits the difference in denaturation and renaturation characteristics of covalently closed circular plasmid DNA and chromosomal DNA. When cells are suspended in a solution of NaOH and sodium dodecyl sulfate, the cells are lysed in high pH (alkaline) environment, additionally, the protein and chromosome DNA become denatured. They precipitate together with cell debris during the centrifugation after the neutralizing solution (acidic potassium acetate) has been added, leaving plasmid DNA in the supernatant. With the alkaline added, plasmid DNA is also denatured, but the two strands remain in close proximity, just as two links of a chain. Upon neutralization, plasmid DNA strands reanneal to their complementary ones and the plasmid is reformed.

This protocol provides an easy and rapid way for plasmid DNA isolation from small-scaled culture or many cell colonies at the same time. The plasmid DNA obtained by this method is of high purity, therefore can be used for enzymatic digestion, PCR, sequencing etc. without further purification.

【Apparatus, materials and reagents】

1. Apparatus

Thermostat incubator, thermostat shaker, atuoclave, table centrifuge, vortex mixer and three pipettes with the capacities of 10 μL, 100 μL and 1 000 μL respectively.

2. Materials

Glucose, trihydroxymethyl aminomethane (Tris), ethylenediamine tetraacetic acid (EDTA), sodium hydroxide (NaOH), sodium dodecyl sulphate (SDS), potassium acetate, glacial acetic acid, chloroform, anhydrous ethanol, RNase, ampicillin, sucrose, bromophenol blue, phenol, 8-quinolinol, β-mercaptoethanol, hydrochloric acid (HCl), *E. coli* strains containing plasmid pUC 19, *Eco*R Ⅰ, isopropanol, 50 mL sterile microcentrifuge tubes, tips, small test tubes.

3. Reagents

(1) Solution Ⅰ (make it fresh just prior to use, store it on ice, heating to room temperature before use): prepare 25 mmol/L Tris · HCl and adjust pH to 8.0, supplemented with 50 mmol/L glucose and 10 mmol/L EDTA (pH 8.0).

(2) Solution Ⅱ: prepare 0.4 mol/L NaOH, supplemented with 2% SDS. Equal volume mixing before use.

(3) Solution Ⅲ: prepare 60 mL 5 mol/L potassium acetate, supplemented with 11.5 mL acetic acid and 28.5 mL ultra-pure water. Store it on ice.

(4) TE buffer: 10 mmol/L Tris · HCl (pH 8.0), supplemented with 1 mmol/L EDTA (pH 8.0)

(5) 70% ethanol (store it in refrigerator at −20 ℃, put it back after use).

(6) RNase solution: dissolve RNase in 10 mmol/L Tris · HCl (pH 7.5) /15 mmol/L NaCl to prepare a 10 mg/mL RNase solution. Heat it at 100 ℃ for 15 mins and then cool it slowly to room temperature. Store it at −20 ℃.

(7) Loading buffer (6 ×): prepare 40% sucrose solution, supplemented with 0.25% bromophenol blue.

(8) Phenol and chloroform (1 : 1) saturated with TE buffer.

【Procedures】

(1) Pick a single colony of bacteria from the LB plate and inoculate it into 100 mL LB culture medium containing corresponding antibiotics. Shake it at 37 ℃ (at about 225 rpm rotation speed) and culture it overnight.

(2) Pipette 1.2 mL of the overnight culture solution into a micro centrifuge tube and centrifuge at 12 000 rpm for 30 seconds.

(3) Remove the supernatant and resuspend the pellet in 100 μL of solution Ⅰ containing 4 mg/mL lysozyme at room temperature. Vortex the cells into suspension, making sure there is no clump of cells.

(4) After a few seconds at room temperature, lysozyme action on the cells will be visible. The suspended cells became slightly emulsus and attached to the tube wall easily during oscillation. Place the tube into an ice bucket and chill for 1 minute.

(5) Add 200 μL of solution Ⅱ into the tube at room temperature and mix it by inversion. Return the tube to the ice bucket cooling for 1 minute.

(6) Add 150 μL of ice-cold solution Ⅲ into the tube and mix it. Return the tube to the ice bucket for 3 minutes. A fluffy white precipitate will come out.

(7) Centrifuge the tube at 12 000 rpm for 2 minutes in a microcentrifuge.

(8) Transfer the supernatant to a fresh tube, and take care to avoid introducing the white sediment.

(9) Add equal volume of the phenol and chloroform solution to the fresh tube. Vortex it for 1 minute and centrifuge it at 12 000 rpm for 2 minutes at room temperature.

(10) Transfer the upper aqueous phase to another fresh tube and add equal volume of chloroform into it. Vortex it for 30 seconds and centrifuge it at 12 000 rpm for 2 minutes.

(11) Transfer the upper aqueous phase to a fresh tube again and add double volume of ice-cold 100% ethanol into it. Mix it and allow precipitating for 5 minutes at −20 ℃.

(12) Centrifuge the tube at 12 000 rpm for 4 minutes or more. Pour off the supernatant, rinse the pellet with ice-cold 70% ethanol and vacuum dry the pellet.

(13) Resuspend the pellet in a fresh tube with 50 μL of TE buffer containing 50 μg/mL RNase. Incubate the tube for 5 minutes at 37 ℃ (If RNase is not used, the DNA fragment smaller than 600 bp will be obscured by RNA when agarose gel electrophoresis).

(14) Label the tube and store it at −20 ℃.

【Result】

Collect the plasmid solution obtained from this experiment in a small tube and label the date and content. Store it at −20 ℃ for further use.

三、聚合酶链式反应（PCR）

【实验目的】

（1）理解聚合酶链式反应的工作原理。

（2）熟悉聚合酶链式反应的操作步骤。

【实验原理】

PCR 技术的基本原理类似于 DNA 的自然复制过程。PCR 技术利用 DNA 聚合酶复制 DNA 的一个特定区域。我们可以选择要扩增的 DNA 的 X 部分，方法是放入短片段的 DNA（引物），与 X 两侧的 DNA 序列杂交，并通过 X 引发 DNA 合成的启动。合成的 X 的两条链以及原始 DNA 链的副本将作为下一轮扩增的模板。这样，所选择的 DNA 区域的数量在每一个循环中不断翻倍，达到起始数量的数百万倍，一直到有足够的 DNA 可通过凝胶电泳看到。

PCR 由变性—退火—延伸三个基本反应步骤构成：①模板 DNA 的变性：模板 DNA 经加热至 93 ℃左右维持一定时间后，模板 DNA 双链或经 PCR 扩增形成的双链 DNA 解离，成为单链，便于与引物结合，为下轮反应做准备；②模板 DNA 与引物的退火（复性）：模板 DNA 经加热变性成单链后，温度降至 55 ℃左右，引物与模板 DNA 单链的互补序列配对结合；③引物的延伸：DNA 模板—引物结合物在 72 ℃、DNA 聚合酶（如 Taq DNA 聚合酶）的作用下，以 dNTP 为反应原料，靶序列为模板，按碱基互补配对与半保留复制原理，合成一条新的与模板 DNA 链互补的半保留复制链，重复循环变性—退火—延伸三过程就可获得更多的“半保留复制链”，而且这种新链又可成为下次循环的模板。每完成一个循环需 2 ~4 分钟，2 ~3 小时就能将待扩目的基因扩增放大几百万倍。

典型的扩增反应包括靶 DNA 样品，热稳定 DNA 聚合酶，两条寡核苷酸引物，dNTPs（各种三磷酸脱氧核苷酸），反应缓冲液，Mg^{2+} 及其他添加物。反应液组分必须进行混合，然后置于可以自动控制不同温度与时间的热循环仪中进行反应。

【实验仪器、材料与试剂】

1．仪器

PCR 热循环仪、琼脂糖凝胶电泳系统、凝胶成像系统。

2．材料

DNA 模板、4 种 dNTPs、正向引物和反向引物、Taq 酶、琼脂糖、DNA 相对分子质量标准物、吸头、小试管。

3．试剂

灭菌水、含 15 mmol/L $MgCl_2$ 10×扩增缓冲液、10 mmol/L dNTPs、50 μmol/L 寡核苷酸引物Ⅰ和引物Ⅱ、5 unit/μL Taq DNA 聚合酶、模板 DNA（1 μg 基因组 DNA，0.1～1 ng 质粒 DNA）。

【实验步骤】

（1）在 PCR 管中将下列各组分混合：

试剂	实验组	对照组
10×PCR 缓冲液	10 μL	10 μL
正向引物	1 μL	1 μL
反向引物	1 μL	1 μL
dNTPs	2 μL	2 μL
模板 DNA	2 μL	0
水	85.5 μL	85.5 μL
Taq DNA 聚合酶	0.5 μL	0.5 μL

（2）将小试管置于热循环仪中预热到 94 ℃（5 分钟）。

（3）按以下程序操作：

①94 ℃，30 秒（预热）；

②55 ℃，30 秒；

③72 ℃，2 分钟；

④从①～③步，30 个循环；

⑤最终于 72 ℃延伸 10 分钟。

（4）在琼脂糖凝胶上准备如下溶液：

①取 3 μL DNA 分子量标准溶液上样。

②取 10 μL 产物加 2 μL 6×载样缓冲液混合上样。

（5）进行琼脂糖凝胶电泳：

①用移液枪将缓冲液与样品或标准品混合，加入胶孔中。

②在 1% 琼脂糖凝胶中进行电泳。当溴酚蓝染料从加样孔中迁移至可以足以将 DNA 片段分开的距离时，关闭电源。

③用 0.5 μg/mL 浓度的溴乙啶溶液染色 10～30 分钟。

④紫外光照射凝胶，然后照相。将产物条带与已知分子量的标准条带进行比较。这样，便可以对合适分子量大小的产物进行鉴定。

【注意事项】

优化聚合酶链式反应须考虑的几个方面：

1. 镁离子浓度

镁离子浓度是一种影响 Taq DNA 聚合酶工作的关键因素。反应组分如模板 DNA、样品中存在的螯合剂（EDTA 或柠檬酸盐）、dNTPs 和蛋白质，都会影响游离镁离子的含量。当镁离子缺少时，Taq DNA 聚合酶会失去活性；当镁离子过量时，会降低聚合酶的忠实性并且会增加非特异性的扩增。因此，根据经验研究每个反应的最佳 $MgCl_2$ 浓度非常重要。为此，需要通过添加 2 μL、3 μL、4 μL、5 μL 25 mmol/L $MgCl_2$ 到 50 μL 反应体系中，以制备一个以 0.5 mmol/L 为增量的含 1.0～3.0 mmol/L Mg^{2+} 的反应系列。

2. 酶浓度

酶浓度的控制，一般在 1.25 单位的 Taq DNA 聚合酶。增加酶量和过度增加延伸时间会增加出现夹带的可能性，5’→3’最终将导致琼脂糖凝胶的 smear 现象。

3. 引物设计

引物设计方面，同样需要注意，PCR 引物长度一般在 15～30 碱基，设计的两引物之间为 DNA 目的区域。引物设计时，其 G+C 含量应在 40%～60% 之间，而且要避免在引物内部出现二级结构。引物的 3’端不应是互补的，以避免在 PCR 反应中产生引物二聚体；也要避免三个 G 或 C 核苷酸在靠近引物 3’端的一排。理想情况下，两种引物应在相同温度下退火。退火温度取决于最低熔化温度（T_m）的引物。估算 T_m 的简单公式：T_m = 20 ℃ ×（A 和 T 残基数）+40 ℃ ×（G 和 C 残基数）。

4. 循环次数

循环次数一般控制在 25 到 30 之间。较多的循环虽然意味着较高的产量，然而各种错误的产物也会增多，一般不能超过 40 个循环。

5. 核酸交叉污染

特别注意，要减少样品之间的交叉污染，防止将一个实验室的 RNA 和 DNA 带入下一个实验室。扩增前和扩增后的工作区以及移液枪应分开使用，应戴手套并经常更换。

【实验结果】

本实验扩增的片段长约 548 bp（如图 3－1）。

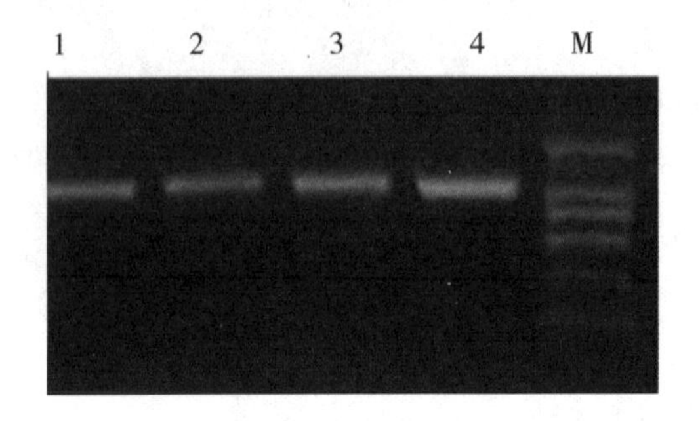

图 3－1　电泳图

注：M，DNA 分子质量标准；泳道 1～4：样品条带。

Ⅲ. Polymerase Chain Reaction

【Objectives】

(1) Understand the principle of PCR.

(2) Be familiar with the procedures of PCR.

【Principle】

The basic principle of PCR technology is similar to the natural replication process of DNA. PCR technology uses DNA polymerase to make a copy of a defined region of DNA. We can select the part X of the DNA we want to amplify by putting in short pieces of DNA (primers) that hybridize to DNA sequences on either side of X and cause initiation of DNA synthesis through X. The copies of both strands of X, as well as the original DNA strand, then serve as templates for the next round of amplification. By this way, the amount of the selected DNA region double over and over with each cycle—up to millions of times the starting amount—until enough DNA can be seen by gel electrophoresis.

PCR consists of three basic reaction steps: denaturation, annealing and extension. ① Denaturation of template DNA. After the template DNA is heated to 93 ℃ for a certain time, the double strands of template DNA or the double strands DNA formed by PCR amplification are dissociated to form a single strand, which is easy to combine with the primer and prepare for the next round of reaction. ② Annealing (refolding) of template DNA and primer. After the template DNA is denatured into single strand by heating, cool it to about 55 ℃, and the primer and complementary sequence of template DNA single strand will be paired and combined. ③ Primer extension. DNA template-primer conjugate, under the action of DNA polymerase (e.g., Taq DNA polymerase) at 72 ℃, with dNTP as reaction material and target sequence as template, a new semiconservative replication chain complementary to template DNA chain will be synthesized according to the principle of base complementary pairing and semi-reserved replication. By repeating the three processes of cyclic denaturation-annealing-extension, more "semiconservative replication chains" will be obtained, and those new chains become templates again for the next cycle. Each cycle takes 2 to 4 minutes. In 2 to 3 hours, the target gene can be amplified millions of times.

A typical amplification reaction includes the sample of target DNA, a thermostable DNA polymerase, two oligonucleotide primers, dNTPs (deoxynucleotide triphosphates), reaction buffer, Mg^{2+} and optimal additives. The components of the reaction are mixed before the reaction is placed in a thermal cycler, which is an automated instrument that takes the reaction through a series of different temperatures for varying amounts of time.

【Apparatus, materials and reagents】

1. Apparatus

PCR thermal cycler, agarose gel electrophoresis system, gel imagining system.

2. Materials

Template DNA, four dNTPs, forward and reverse primers, Taq polymerase, agarose, DNA marker, tips and mini tubes.

3. Reagents

Sterilized water, 10 × amplification buffer with 15 mmol/L $MgCl_2$, 10 mmol/L dNTPs, 50 μmol/L oligonucleotide primers Ⅰ and Ⅱ, 5 unit/μL Taq DNA polymerase and DNA template (1 μg genome DNA, 0.1 – 1 ng plasmid DNA).

【Procedures】

(1) Add the following solutions for each reaction in a PCR tube:

Reagents	Experimental group	Control group
10 × PCR buffer	10 μL	10 μL
forward primer	1 μL	1 μL
reverse primer	1 μL	1 μL
dNTPs	2 μL	2 μL
template DNA	2 μL	0
sterile water	85.5 μL	85.5 μL
Taq DNA polymerase	0.5 μL	0.5 μL

(2) Place the tubes in a thermal cycler preheated to 94 ℃ (5 minutes).

(3) Perform the following procedures:

①94 ℃ for 30 seconds (preheat);

②55 ℃ for 30 seconds;

③72 ℃ for 2 minutes;

④From step ① to ③ makes 30 cycles;

⑤72 ℃ for 10 minutes in final extension.

(4) Prepare the following solutions on an agarose gel:

①3 μL DNA molecular weight markers.

②Mixed solution of 10 μL PCR product and 2 μL 6 × loading buffer .

(5) Perform the following procedures:

①Load samples or marker mixed with loading buffer into wells with a pipette.

②Electrophorese on a 1% agarose gel. Turn off the power supply when bromphenol blue dye from the loading buffer has migrated a distance sufficient for separation of DNA fragments.

③Stain with 0.5 μg/mL ethidium bromide solution for 10 to 30 minutes.

④Illuminate the gel with UV light and then photograph. By comparing product bands with bands from the known molecular-weight marker, PCR products with appropriate molecular weight can be identified.

【Notes】

There are several points need to be considered for PCR optimization:

1. Magnesium concentration

Magnesium concentration is a crucial factor affecting the performance of Taq DNA polymerase. Reaction components, including template DNA, chelating agents present in the sample (e.g., EDTA or citrate), dNTPs and proteins, all affect the amount of free magnesium. In the absence of adequate free magnesium, Taq DNA polymerase is inactive. Conversely, excess free magnesium reduces enzyme fidelity and increases the level of non-specific amplification. For these reasons, it is important to empirically determine the optimal $MgCl_2$ concentration for each reaction. Therefore, we can prepare a reaction series containing 1.0 – 3.0 mmol/L Mg^{2+} in 0.5 mmol/L increments by adding 2, 3, 4 or 5 μL of 25 mmol/L $MgCl_2$ stock to 50 μL reactions.

2. Enzymes concentration

It is recommended that using 1.25 units of Taq DNA polymerase controls the concentration of enzymes. It should also be noted that increasing amounts of enzyme and excessively long extension times will increase the likelihood of generating artifacts associated with the intrinsic 5' →3' exonuclease activity associated with Taq DNA polymerase, resulting in smearing in agarose gels.

3. Primer design

PCR primers generally range in length from 15 – 30 bases and are designed to flank the target region. Primers should contain 40% – 60% G + C and care should be taken to avoid sequences which would produce internal secondary structure. The 3' -ends of the primers should not be complementary to avoid the production of primer-dimmers in the PCR reaction. Also, avoid three G or C nucleotides in a row near the 3' -ends of the primer. Ideally, both primers should anneal at the same temperature. The annealing temperature depends on the primer with the lowest melting temperature (T_m). A simple formula for estimating T_m: $T_m = 20\ ℃ \times$ (number of A and T residues) + $40\ ℃ \times$ (number of G and C residues).

4. Number of cycles

The number of cycles is usually between 25 and 30. Although more cycles indicate a greater yield of product, the greater probability of generating various artifacts (e. g., mispriming products) occurs due to increasing number of cycles. Cycle should be no more than 40 cycles usually.

5. Nucleic acid cross-contamination

It is important to take great care to avoid cross-contamination between samples and to prevent carryover of RNA and DNA from one experiment to the next. Use separate work areas and pipettes for pre- and post-amplification steps. Wear gloves and change them often.

【Result】

The length of the amplified fragment is about 548 bp (Figure Ⅲ – 1).

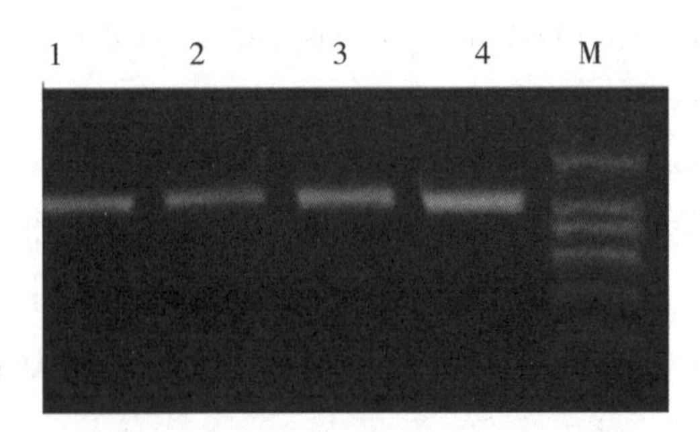

Figure Ⅲ – 1 Gel electrophoresis image

Note: M, DNA markers; lane 1 – 4: sample bands.

四、质粒 DNA 的酶切及凝胶电泳

【实验目的】

（1）理解限制性内切酶的工作原理。

（2）掌握琼脂糖凝胶电泳的原理和操作方法。

【实验原理】

限制性酶在脱氧核糖和磷酸基因团间对 DNA 进行水解，结果分别在两条双链的末端留下 5’ 端磷酸团和 3’ 端羟基。尽管其切割效率很低，仍然有少数限制性酶可以切割单链 DNA。

核酸限制性内切酶的一个特性是其所识别的序列是回文结构，即两条链在 5’→3’方向上分别具有相同的序列。典型的限制性酶切位点（Ⅱ型）是一个由 4 ~ 8 个 bp 组成的具有一个旋转对称轴的精确回文结构（如 *EcoR* Ⅰ的识别序列为 GAATTC）。一些限制性酶在 DNA 双链识别中心周围做交错对称切割，另一些酶在识别位点中心做对称切割。前一种情况产生 DNA 黏性末端，后一种情况产生 DNA 钝端。

在进行限制性酶消化时，将双链 DNA 分子与适量的限制性酶放进相应的缓冲液中进行混合保温，并在该酶的最适温度下进行反应。典型的消化反应中每毫克起始 DNA 有一个单位的酶，一个酶单位通常定义为 1 小时内在合适的温度下，一般是 37 ℃下，用于完全消化 1 毫克双链 DNA 的酶数量。为了保证消化完全，通常需要保温 1 ~ 3 小时。

电泳采用的是琼脂糖。琼脂糖是一种海藻多糖，琼脂糖胶分离范围很大，但其分辨率却相对较低。通过改变琼脂糖凝胶的浓度，应用标准的电泳技术可以分离 200 bp 到 50 000 bp 大小的 DNA 片段。琼脂糖凝胶浓度越高，凝胶就越硬。较高浓度的琼脂糖凝胶有利于较小的 DNA 的分离，而较低浓度的琼脂糖凝胶则可以分离较大的 DNA 片段。

琼脂糖凝胶浓度一般在 0.5% 与 2% 之间。通过观察示踪染料的迁移距离可以判断 DNA 的迁移距离。溴酚蓝和二甲苯青染料在琼脂糖凝胶中的迁移速率大致分别与 300 bp 和 400 bp 大小的双链 DNA 片段相同。迁移足够距离后，就可以通过溴化乙啶染色来观察 DNA 片段。溴化乙啶是一种荧光染料，它嵌插在 DNA 和 RNA 碱基之间。它可以在制作凝胶时混入其中，使其在电泳时进行染色，也可以待电泳完成后将凝胶浸泡在稀释的溴化乙啶溶液中进行染色。必须将凝胶置于紫外分析检测器中才可以对凝胶中的 DNA 或 RNA 进行观察。

【实验仪器、材料与试剂】

1. 仪器

恒温培养箱、琼脂糖凝胶电泳系统、台式离心机、高压灭菌锅、恒温水浴锅、凝胶成像系统、移液枪（10 μL、100 μL、1 000 μL 各一支）。

2. 材料

未线性化的质粒 pUC 19、限制性酶 *Eco*R Ⅰ、DNA 分子标准、三羟甲基氨基甲烷（Tris）、硼酸、乙二胺四乙酸（EDTA）、溴酚蓝、二甲苯蓝（FF）、蔗糖、琼脂糖、溴化乙啶、吸头、小试管。

3. 试剂

（1）限制性酶缓冲液。

10 × *Eco*R Ⅰ缓冲液［含 100 mmol/L Tris－HCl（pH 7.5，温度 25 ℃），50 mmol/L NaCl 和10 mmol/L 硫酸镁与 0.025% Triton X－100）］。

（2）TAE（Tris－乙酸，EDTA）（浓缩储备溶液 50 ×）。

取 242 g Tris 碱，57.1 mL 冰醋酸和 100 mL 0.5 mol/L EDTA（pH 8.0）加入 600 mL ddH_2O中，剧烈搅拌，加入 ddH_2O 至 1 000 mL。

（3）溴化乙啶原液（10 mg/mL）。

在 100 mL 水中加入 1 g 溴化乙啶，用磁力搅拌器搅拌数小时，移入深色瓶中，4 ℃保存（溴化乙啶是一种强诱变剂，称重时应戴上手套和口罩。如果接触，立即用大量水冲洗）。

（4）TBE（5 倍体积的 TBE 贮存液 1 L）。

取 54 g Tris、27.5 g 硼酸，20 mL 0.5 mol/L EDTA，调节 pH 至 8.0，补充 ddH_2O 至1 000 mL。

（5）凝胶上样缓冲液（6 ×）。

取 40 g 蔗糖、0.25 g 溴酚蓝和 0.25 g 二甲苯蓝（FF），补水至 100 mL。

（6）1 % 琼脂糖和 0.5 μg/mL 溴化乙啶。

【实验方法】

（1）－20 ℃冰箱中取出实验三制备的质粒样品，让其升至室温。

（2）限制性酶消化是在特定的缓冲液中进行的。缓冲液以 10 × 浓缩贮存液供应，使用时必须稀释成 1 × 反应液。反应可以在 20 μL 总体积中进行。

（3）在标有“E”的试管中加入 10 μL 质粒 DNA，7 μL ddH_2O，2 μL *Eco*R Ⅰ 10 × 缓

冲液，1 μL *Eco*R Ⅰ酶液。

（4）加入所有组分后，盖上盖子，轻弹混匀，然后在微量离心机上离心 2～3 秒，将试剂甩至管底。

（5）将试管于 37 ℃水浴保温 2 小时。

（6）制备电泳用的凝胶。

①搭好凝胶电泳槽。梳齿应该保持笔直，在梳齿底部与胶槽之间应保持几毫米的间隙。

②称取适量琼脂糖，加入 Tris－乙酸缓冲液中，用微波炉加热至溶解（例如称取 1 g 琼脂糖，溶于 100 mL Tris－乙酸缓冲液中，制成 1 % 的琼脂糖）。

③将琼脂糖溶液缓缓倒入插有齿梳的凝胶板（避免产生气泡）。如果有气泡，用巴斯德吸管去除。根据凝胶大小，让其凝固 20 分钟至 60 分钟。

④在琼脂糖凝固过程中，准备好 DNA 样品，且向其中加入含有染料如溴酚蓝的加样缓冲液。

⑤琼脂糖凝胶凝固之后，移走齿梳，将凝胶板移至电泳槽上，加样孔在靠近阴极的一端（黑色端）。向槽中加入适量的 TAE 缓冲液，通常应没过胶面 1 cm。再加入溴化乙啶，混合完全（其最终浓度达到 1 μg/mL）。另一种方法是将溴化乙啶加入凝胶中；用微波炉将琼脂糖煮沸，然后冷却至 50 ℃，加入溴化乙啶。

（7）将 10 μL 反应样品与 2 μL 载样染料混匀准备点样。

（8）按以下顺序上样 10 μL（标准分子 5 μL），并在笔记本上记录下加样顺序。

泳道	样品
A	DNA 大小标准
B	*Eco*R Ⅰ限制性消化的样品
C	未切割的质粒

（9）以 10 V/cm 的电压进行电泳直至深蓝染料到达凝胶末端（约 1 小时）。

（10）电泳完拍照。如果凝胶没有结果，则可能需要重新消化或重新电泳，再将限制性酶切质粒与未切割质粒进行比较。

（11）测定质粒片段的大小。只有线性的（可消化的）质粒 DNA 分子与凝胶中质粒的大小成反比，因此我们决不能根据迁移距离来确定没有切割的质粒的大小。

【实验结果】

质粒 DNA 酶切后的电泳结果如图 4－1 所示。

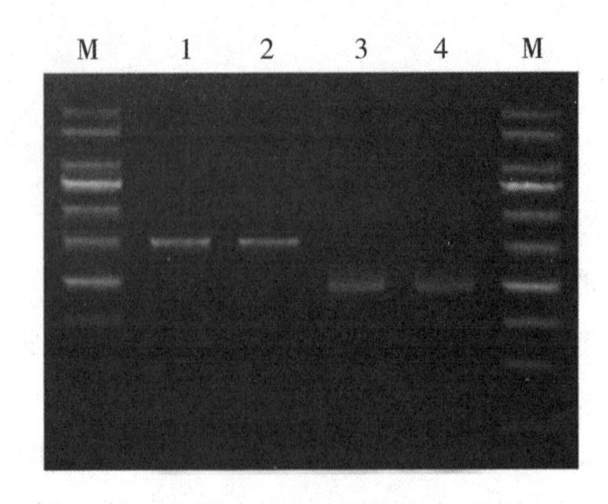

图4－1　质粒 pUC 19 及其酶切产物电泳图

注：泳道 M，分子质量标准；泳道 1 和 2，质粒 pUC 19；泳道 3 和 4，质粒 pUC 19 的 *Eco*R Ⅰ 单酶切产物。

【注意事项】

（1）比较用酶消化质粒的通道和含有未切割 DNA 的通道，其结果应该不一样。如果没有不同，则该消化过程可能没有发生。

（2）测定酶消化质粒的通道中片段的大小，只有线性（消化的）DNA 通过凝胶的速度与碱基对的大小成反比。因此，不能根据未切割的质粒 DNA 的迁移来确定质粒的大小。

（3）用移液枪缓慢将 DNA 样品垂直加入样孔。加完所有样品后，将电泳槽与电源正确连接（黑色对阴极，红色对阳极）。打开电源之前要调好电压（我们采用 110 伏的电源）和时间。观察铂丝是否已连接到加样孔附近的黑色端以确保导线连接正确。如果导线连接正确，会有气泡缓慢上升。

（4）电泳过程需要一定的时间，其长短取决于凝胶的长度及使用的电压大小。凝胶越长电压越低，则所需时间就越长。但是，在分离较大的 DNA 片段时，使用高电压效果不好。

（5）未切割质粒 DNA 在其泳道上也许会出现几个条带，之所以这样是因为质粒 DNA 在琼脂糖凝胶中的迁移距离是由其分子构象及其碱基对大小所决定的。质粒 DNA 可以下列三种主要构象中的任何一种形式存在：

①超螺旋：尽管质粒通常以开环的形式进行描述，然而在细菌细胞内 DNA 链却是盘绕在组蛋白样的蛋白质周围形成一种致密的结构。这就是所谓超螺旋结构，由于其结构致密，它在凝胶中的泳动速度最快。

②切口：在质粒 DNA 复制过程中，拓扑异构酶Ⅰ会在 DNA 双螺旋中的一条链中引入一个切口，解开质粒的超螺旋。在质粒分离过程中由于物理剪切和酶的切割作用同样也会在超螺旋质粒中引入切口从而产生松散的开环结构。这种形式的质粒迁移速率最慢，其“松散”的分子形式阻碍了它在琼脂糖凝胶中的运动。

③线性：当 DNA 损伤在 DNA 双链相对应的两条链上同时产生切口时，就会出现线性质粒 DNA。这种 DNA 的泳动速率介于超螺旋与切口质粒 DNA 之间。质粒制备过程中出现线性 DNA 说明存在核酸酶污染或实验操作有问题。

【小知识】

（1）TAE 和 TBE 都是常用的缓冲液，TBE 具有比 TAE 更高的缓冲能力。

（2）负载染料溴酚蓝与 DNA 的迁移量约为 0.5 kb，为片段的最快迁移率提供了一个指标。

（3）DNA 的迁移速率取决于以下因素：

①DNA 的分子大小：分子越小迁移越快。

②琼脂糖的浓度：浓度越低迁移越快。

③DNA 的构象：环状的或带切口的环状 DNA 通常比线状的 DNA 迁移要快。

④两个电极之间的单位厘米之间的电压：电压越高，迁移越快。

（4）可能出现的问题。

如果 DNA 条带不清晰且不均匀，可能是由于以下原因：

①DNA 超载。

②电压过高。

③凝胶孔被破坏了。

④凝胶里有气泡。

Ⅳ. Digestion and Agarose Gel Electrophoresis of Plasmid DNA

【Objectives】

(1) Understand the working principle of restriction enzyme.

(2) Master the principle and operation method of agarose gel electrophoresis.

【Principle】

Restriction enzymes hydrolyze the backbone of DNA between deoxyribose and phosphate groups. This leaves a phosphate group on the 5' ends and a hydroxyl on the 3' ends of both strands. A few restriction enzymes can cleave single-stranded DNA, although usually at low efficiency.

One unique feature of restriction enzymes is that the nucleotide sequences they recognize are palindromic. That means, the two strands have the same sequence at the direction 5' →3', respectively. The typical restriction enzyme site (type Ⅱ) is an exact palindrome of 4 – 8 bp with an axis of rotational symmetry (e. g. , *Eco*R Ⅰ recognition sequence is GAATTC) . Some restriction endonucleases make staggered symmetrical cuts away from the center of their recognition site within the DNA duplex; while others make symmetrical cuts in the middle of their recognition site. Enzymes that make staggered cuts leave the resultant DNA with cohesive or sticky ends. Enzymes that cleave the DNA at the center of the recognition sequence leave blunt-ended fragments of DNA.

Restriction enzyme digestion is performed by incubating double-stranded DNA molecules with an appropriate amount of restriction enzyme, in its respective buffer as recommended by the supplier, and at the optimal temperature for that specific enzyme. Typical digestion included a unit of enzyme per microgram of starting DNA, and one enzyme unit usually (depending on the supplier) is defined as the amount of enzyme needed to completely digest one microgram of double-stranded DNA in one hour at the appropriate temperature, typically 37 ℃. To insure complete digestion, the reactions are usually incubated for 1 – 3 hours.

Agarose is applied to perform gel electrophoresis of DNA. Agarose is a polysaccharide from seaweed. Agarose gel has a large range of separation, but relatively low resolution. By varying the concentration of agarose, fragments of DNA from about 200 to 50 000 bp can be separated using standard electrophoretic techniques. The higher the agarose concentration, the "stiffer" the gel is. Higher concentration of agarose facilitate separation of smaller DNAs, while the lower ones allow resolution of larger DNAs.

The agarose gel is typically used at a concentration from 0.5% to 2%. The distance DNA has migrated in the gel can be judged by visually monitoring migration of the tracking dyes. Bromophenol blue and xylene cyanol dyes migrate through agarose gel at roughly the same rate as double-stranded DNA fragments of 300 and 400 bp respectively. When adequate migration has occurred, DNA fragments can be visualized by straining with ethidium bromide. The fluorescent dye will be embedd in DNA and RNA bases. It is often incorporated into the gel so that staining occurs during electrophoresis, but the gel can also be strained after electrophoresis by soaking in a dilute solution of ethidium bromide. To visualize DNA or RNA, the gel should be placed on an ultraviolet transilluminator.

【Apparatus, materials and reagents】

1. Apparatus

Thermostat incubator, agarose gel electrophoresis system, table centrifuge, autoclave, constant temperature incubator, gel imaging system and pipettes with the capacities of 10 μL, 100 μL and 1 000 μL respectively.

2. Materials

Unlinearized plasmid pUC 19, *Eco*R Ⅰ, DNA markers, trihydroxymethyl aminomethane (Tris), boric acid, ethylenediamine tetraacetic acid (EDTA), bromophenol blue, xylene blue (FF), sucrose, agarose, ethidium bromide, tips and mini tubes.

3. Reagents

(1) Buffers for restriction enzymes.

10 × buffer for *Eco*R Ⅰ containing 100 mmol/L Tris-HCl (pH 7.5 at 25 ℃), 50 mmol/L NaCl, 10 mmol/L magnesium chloride and 0.025% Triton X-100.

(2) TAE (Tris-Acetate, EDTA) (concentrated stock solution 50 ×).

Add 242 g Tris base, 57.1 mL glacial acetic acid and 100 mL 0.5 mol/L EDTA (pH 8.0) into 600 mL ddH_2O, stir vigorously, and add ddH_2O to 1 000 mL.

(3) Ethidium bromide stock (10 mg/mL).

Add 1 g of ethidium bromide to 100 mL of H_2O, stir it with a magnetic stirrer for several hours. The solution should be transferred to a dark bottle and stored at 4 ℃ (Ethidium bromide is a powerful mutagen. Wear gloves and a mask when weighing it out. In case of contact, flush immediately with copious amounts of water) .

(4) TBE (5 × stock solution, 1 L).

Add 54 g Tris base, 27.5 g boric acid and 20 mL 0.5 mol/L EDTA into a volumetric flask, adjust pH to 8.0 and add ddH_2O to 1 L.

(5) Loading buffer (6 ×) .

Dissolve 0.25 g bromophenol blue, 0.25 g xylene cyanol (FF) and 40 g sucrose , add ddH_2O to 100 mL.

(6) 1% agarose gel and 0.5 μg /mL ethidium bromide.

【Procedures】

(1) Take out the plasmid sample prepared in Experiment Ⅲ from the refrigerator at −20 ℃ and leave it to room temperature.

(2) Restriction enzyme *Eco*R Ⅰ digestion is performed in a special buffer for *Eco*R Ⅰ. The buffer is a 10 × concentrated storage solution. It must be diluted to 1 × reaction solution. The reaction can be performed within a total volume of 20 μL.

(3) Add 10 μL of plasmid DNA, 7 μL of ddH_2O, 2 μL of 10 × buffer for *Eco*R Ⅰ, and 1 μL of *Eco*R Ⅰ into the tube marked with "E" .

(4) After adding all ingredients, cap the tubes, flick them to mix the contents, then pool the reagents in the bottom by spinning the tubes in the microcentrifuge for 2 to 3 seconds.

(5) Incubate the tubes in a 37 ℃ water bath for 2 hours.

(6) Prepare the gel.

①Set up the gel apparatus as indicated. The comb should be straight, and there should be a few millimeters of clearance between the bottom of the comb and the bottom of the gel tray.

②Weigh proper amount of agarose (e.g., 1.0 g agarose in 100 mL TAE buffer equals 1% agarose); dissolve it in the TAE buffer by heating it in a microwave oven.

③Pour the agarose solution slowly (avoid bubbles) onto a gel bed with the comb inserted. Use a pipette to remove the bubble if there is any. Let gel polymerize for 20 to 60 minutes, depending upon the gel size.

④While the agarose gel is polymerizing, prepare the DNA sample and mix proper amount of DNA with the loading buffer containing dyes such as bromophenol blue.

⑤Remove the comb after the agarose gel has polymerized and place the gel bed onto the electrophoresis tank with the wells near the cathode (black terminal) . Fill the tank with proper amount of TAE buffer. Usually the buffer is about 1 cm above the gel. Add ethidium bromide (final concentration is 1 μg/mL) to the TAE buffer and mix it well. Alternatively, ethidium bromide can be added to the agarose gel. Boil the agarose in the microwave and then cool it to about 50 ℃, finally add ethidium bromide.

(7) Prepare the samples for electrophoresis and mix 10 μL of the samples with 2 μL of loading dye.

(8) Load 10 μL of the samples (5 μL of markers) onto the gel in the following order and record the order of your samples in your notebook.

Lane	Sample
A	DNA size markers
B	Restriction digestion *Eco*R Ⅰ
C	Uncut plasmid

(9) Perform gel electrophoresis at 10 V/cm until the dark blue dye has reached the end of the gel (about 1 hour).

(10) Take a picture of the finished gel. If the gel does not perform the result, it may need to be re-digested or re-electrophoretic. Compare the lane of restriction enzyme cut plasmid with that of the uncut plasmid.

(11) Determine the size of plasmid fragments in the plasmids. Only the linear (digested) plasmid DNA molecules are inversely proportional to the size of the plasmid in the gel, so we should never determine the size of the plasmid that is not cut according to the migration distance.

【Result】

The electrophoresis results of plasmid DNA after enzyme digestion are shown in Figure Ⅳ-1.

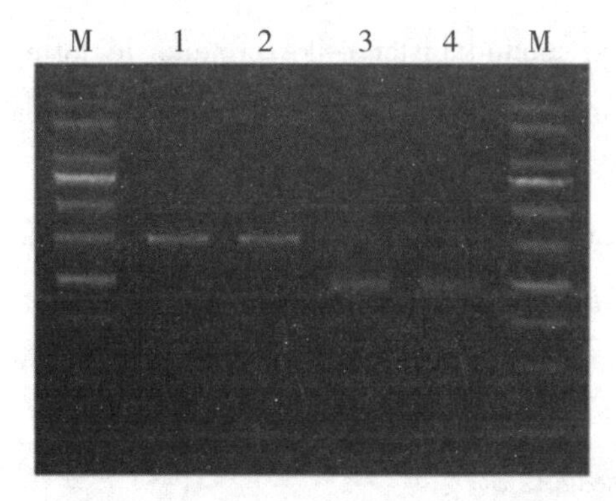

Figure Ⅳ-1 Electrophoretic map of plasmid pUC 19 and its digested products

Note: Lane M, molecular weight standard; lane 1 and 2, plasmid pUC 19; lane 3 and 4, *Eco*R Ⅰ single digested product of plasmid pUC 19.

【Notes】

(1) Compare the lanes in which you digested the plasmid with enzymes to the lane containing the uncut DNA. They should look different. If they do not, the digestion probably did not work.

(2) Determine the size of the fragments in the lanes in which the plasmid was digested with enzyme. Only linear (digested) DNA moves through the gel at a rate that is inversely proportional to its size in base pairs. Therefore, we cannot determine the size of a plasmid based on the migration of uncut plasmid DNA.

(3) Slowly add the DNA sample vertically into the sample well with a pipette. After all the samples have been loaded to the wells, connect the gel tank with the power supply properly (black to cathode [−] and red to anode [+]) . Set the voltage (the voltage we use is 110 V) and time before turning on the power supply. Make sure the leads have been properly connected by confirming the platinum wire at the black terminal near the wells. Bubbles will slowly riss if the leads have a good connection.

(4) Allow electrophoresis to progress for appropriate time. The timing of electrophoresis depends on the length of the gel and the strength of voltage applied. The longer the gel and the lower the voltage, the longer time is needed. However, high voltages are significantly less effective in case of resolving large DNA fragments.

(5) The uncut DNA lane may have several bands in it. This occurs because the migration of plasmid DNA in an agarose gel depends on its molecular conformation as well as its size in base pairs. Plasmid DNA can exist in any one of the following three major conformations:

①Supercoiled. Although a plasmid is usually pictured as an open circle, within a bacterial cell the DNA strand is coiled around histone-like proteins to form a compact structure. This is called supercoiling and this form of the plasmid will move the fastest through the gel due to its compact structure.

②Nicked. During plasmid DNA replication, topoisomerase Ⅰ introduces a nick into one strand of the DNA helix and uncoils the plasmid. Physical sheeting and enzymatic cleavage during plasmid isolation may also introduce nicks into the supercoiled plasmid to produce a relaxing open circular structure. This form is the slowest migrating one of plasmid. Its "floppy" shape impedes movement through the agarose gel.

③Linear. Linear plasmid DNA occurs when damage results in strand nicks directly opposite each other on the DNA helix. This DNA moves at an intermediate rate between supercoiled and nicked plasmid DNA. The presence of linear DNA in a plasmid preparation is a sign of either nuclease contamination or sloppy lab procedure.

【Tips】

(1) Both TAE and TBE are common buffers. TBE has a relatively higher buffering capacity than TAE.

(2) The loading dye bromophenol blue migrates with DNA at the amount of about 0. 5 kb and provides an index of the mobility of the fastest fragments.

(3) The migration of the DNA depends on the following factors:

①Molecular size of DNA: the smaller the DNA is, the faster the migration.

②Concentration of agarose: the lower the concentration is, the faster the migration.

③DNA conformation: circular or nicked DNA often migrate faster than that of linear DNA.

④Voltage between electrodes per cm: the higher the voltage is, the faster the migration.

(4) Problems may occur.

If the DNA bands are not sharp and uniform, it may be due to the following reasons:

①Overloaded DNA.

②Overhigh voltage.

③The destroyed gel pore.

④Bubble in gel.

五、重组质粒的连接

【实验目的】

学习掌握 DNA 连接和 TA 克隆的技术。

【实验原理】

线性 DNA 片段以共价键连在一起的过程称为连接。更明确地说，连接包括在一个核苷的3’端羟基与另外一个核苷酸的5’端磷酸基团之间形成磷酸二酯键的过程。一般采用 T_4 DNA 连接酶连接 DNA 片段，它来源于 T_4噬菌体。T_4 DNA 连接酶可以连接具有单链突出的黏性末端片段，但同时也可以连接具有平齐末端的片段，只是此时通常需要较高浓度的连接酶。除了水之外连接反应通常还需要三种组分：两个或两个以上的具有匹配的黏性末端或平齐末端的 DNA 片段，含有 ATP 的缓冲液和 T_4 DNA 连接酶。

TA 克隆是一种一步就可直接将 PCR 产物质粒载体快速克隆的方法。TA 克隆利用的是在 PCR 反应中热稳定聚合酶（即 Taq 聚合酶）可以向 PCR 产物 3’末端不依赖模板而加上一个腺苷酸（A）的特性。这些末端突出的3’A 即可用于将 PCR 产物直接插入到插入位点具有一个3’T 突出的线性 T 载体分子。这一方法省略了诸如用 Klenow 酶或 T_4多聚激酶对 PCR 产物进行修饰以使其形成钝端或添加插入用的接头等一些费时的酶法修饰过程。

【实验仪器、材料与试剂】

1. 仪器

恒温培养箱、移液枪（10 μL、100 μL、1 000 μL 各一支）。

2. 材料

具有 T 突出的 T－载体（例如 10GEM－T 载体，Promega 公司出品）、PCR 产物、T_4 DNA 连接酶、吸头、小试管。

3. 试剂

T_4 DNA 连接酶缓冲液。

【实验步骤】

（1）取一个微量离心管，添加以下连接混合物：

25 μg/mL 扩增的靶 DNA	1 μL
具有 T 尾巴的质粒	20 ng
10×连接缓冲液	1 μL
T_4 DNA 连接酶（3 Weiss 单位/μL）	1 μL
加 ddH_2O 至 10 μL 终体积	

如果需要的话可添加 ATP 至 1 mmol/L 终浓度。载体与扩增的 DNA 片段的摩尔比以 1∶5为宜。准备一个对照组，对照组除了不含有扩增的靶 DNA 外，含有以上所有列出的反应物。

（2）将连接混合物于 16 ℃连接 12 小时。

【实验结果】

连接好的产物用于后续实验。

【注意事项】

（1）任何条件下的 PCR 产物均可用于 TA 克隆，但由于克隆效率极低，不应使用具有广泛 3’至 5’核酸外切酶活性的聚合酶（如 Vent 和 pfu）。除非 PCR 产物含有多个物种，否则不需要清洗或纯化 PCR 产物。

（2）T_4 DNA 连接酶 10×缓冲液中含有 ATP，在温度波动时会降解。为避免多次冷冻和解冻，应一次性使用小份缓冲液。

（3）如果形成沉淀物，涡旋解冻的 T_4 DNA 连接酶 10×缓冲液，直到沉淀物回到溶液中（涡旋前，在手指之间滚动试管，有助于加热溶液）。

（4）低温连接对悬臂单碱基的退火是必须的。连接温度高于 15 ℃可显著减少重组体的数量。

V. Ligation of DNA and TA Cloning

【Objective】

Learn the method for ligation of DNA and TA Cloning.

【Principle】

The process, joining linear DNA fragments together with covalent bonds, is called ligation. More specifically, DNA ligation involves creating a phosphodiester bond between the 3' hydroxyl of one nucleotide and the 5' phosphate of another. The enzyme used to ligate DNA fragments is T_4 DNA ligase, which originates from the T_4 bacteriophage. This enzyme ligates DNA fragments having overhanging, cohesive ends. T_4 DNA ligase also ligates fragments with blunt ends, although higher concentrations of the enzyme are usually recommended for this purpose. A ligation reaction requires three ingredients in addition to water: two or more fragments of DNA that have either compatible cohesive ("sticky") or blunt ends, a buffer which contains ATP and T_4 DNA ligase.

TA Cloning is a quick, one-step method for the direct insertion of a PCR product into a plasmid vector. TA Cloning takes advantage of the template independent addition of a single adenosine (A) to the 3' -ends of PCR products by activity of thermostable polymerases (i. e. ,Taq polymerase) used in PCR. These 3' A- overhangs are used to insert the PCR product directly into the T vector which is a linear molecule and contains single 3' T-overhangs at its insertion site. This procedure eliminates time consuming enzymatic modifications of the PCR product such as treating with Klenow or T_4 polymerase to create blunt ends or adding adapter for insertion.

【Apparatus, materials and reagents】

1. Apparatus

Thermostat incubator and pipettes with the capacities of 10 μL, 100 μL and 1 000 μL respectively.

2. Materials

T-vector with T overhang (such as 10GEM-T vector, Promega Corporation), PCR product,

T_4 DNA ligase, tips and mini tubes.

3. Reagents

Buffer for T_4 DNA ligase.

【Procedures】

(1) In a microfuge tube, set up the following ligation mixture:

25 μg/mL amplified target DNA	1 μL
T-tailed plasmid	20 ng
10 × ligation buffer	1 μL
T_4 DNA ligase (3 Weiss units/μL)	1 μL

Add ddH_2O to a final volume of 10 μL

If necessary, add ATP to a final concentration of 1 mmol/L. A 1 : 5 molar ratio of vector to amplified DNA fragment is recommended. Set up a control reaction that contains all the reagents listed above except the amplified target DNA.

(2) Incubate the ligation mixture for 12 hours at 16 ℃.

【Result】

Obtain the connected product for later usage.

【Notes】

(1) PCR products under any conditions may be used for TA Cloning but polymerases with extensive 3' to 5' exonuclease activity (such as Vent and pfu) should not be used due to extremely low cloning efficiencies. There is no need to clean or purify PCR product unless the products contain multiple species.

(2) T_4 DNA ligase 10 × buffer contains ATP which degrades during temperature fluctuations. Avoid multiple freezing and thawing by making single-use aliquots of the buffer.

(3) If precipitation is formed, vortex the thawed T_4 DNA ligase 10 × buffer until the precipitate is back in solution (It may help to heat the solution by rolling the tube between your fingers before vortexing).

(4) Low temperature ligation is necessary for annealing of single base overhangs. Ligation temperatures higher than 15 ℃ may significantly reduce the number of recombinants.

六、感受态大肠杆菌的制备、重组体的转化及筛选

【实验目的】

（1）学习掌握氯化钙制备大肠杆菌感受态细胞以及质粒转化的方法。

（2）理解通过 α 互补筛选菌落的原理和流程。

【实验原理】

细菌细胞由于吸入纯化的 DNA 而发生转化，转化也是将重组 DNA 分子导入大肠杆菌最常用的手段。然而，DNA 在通常情况下并不能穿过大多数细菌的细胞膜。为了将 DNA 透过细胞膜导入细胞，细菌的细胞必须处于感受态。就大肠杆菌而言，感受态细胞可以在加入 DNA 之前用氯化钙处理诱导而成。Ca^{2+} 使细胞膜失去稳定性，并与随后加入的 DNA 形成磷酸钙－DNA 复合物，黏附在细胞表面；当细胞进行短暂 42 ℃热休克处理时，便会吸入 DNA。冰冷氯化钙处理可诱导细菌产生感受态，而热休克则可以打开细胞膜让 DNA 进入细胞。

本实验所介绍的方法简单快速，每毫克超螺旋质粒 DNA 可以产生多达 10^7 个转化子。此后该技术又做了许多改进，目的都是提高转化效率。转化效率是指每 1 微克 DNA 转化所得到的转化细胞数。实际上所使用的 DNA 数量要小得多（通常为 5 ~ 100 ng），因为过多的 DNA 会抑制转化过程。

当大肠杆菌中 β－半乳糖苷酶的两个无活性片段结合在一起形成具有功能的酶就产生 α 互补。许多质粒载体携有 β－半乳糖苷酶基因前 146 个氨基酸残基编码信息的 DNA 短片段。这类载体一般与表达 β－半乳糖苷酶酸羧基末端部分的宿主菌配套使用。尽管宿主菌与质粒编码的 β－半乳糖苷酶片段各自都没有酶活性，但它们结合成一体，形成具有酶活性的蛋白。由于产生 α 互补而产生的 Lac^+ 的细菌很容易进行识别，因为它们在有显色底物 X－gal（5 溴－4 氯－3 吲－ β－半乳糖苷）存在时会形成蓝色。然而，在质粒的多克隆位点插入外源 DNA 片段后，几乎不可避免地导致产生无 α 互补能力的氨基端片段。因此，载有重组质粒的细菌就不能水解 X－gal 并形成白色菌落。这一切简单的蓝—白色试验方法的使用很大程度上简化了质粒载体重组子的鉴定过程。

【实验仪器、材料与试剂】

1. 仪器

恒温摇床、恒温箱、恒温水浴锅、低温离心机、超净工作台、－70 ℃冰箱、移液枪

（10 μL、100 μL、1 000 μL 各一支）。

2. 材料

氯化钙（$CaCl_2$）、胰蛋白胨、酵母提取物、氯化钠（NaCl）、氨苄西林、大肠杆菌 $DH_{5\alpha}$、pUC 19 质粒、50 mL 离心管、吸头、小试管、试管、培养皿、锥形瓶等。

3. 试剂

0.1 mol/L $CaCl_2$溶液、LB 液体培养基：配制每升培养基，应在 950 mL 去离子水中加入：10 g 胰蛋白胨（bactol-tryptone），0.5 g 酵母提取物（bactol-yeast extract），0.5 g NaCl，摇动容器直至溶质完全溶解，用 NaOH 调节 pH 至 7.0（每升加 5 mol/L NaOH 200 μL），加入去离子水至总体积为 1 L，121 ℃湿热灭菌 20 分钟。氨苄西林（Amp），用无菌水配制成 100 mg/mL 溶液，置 -20℃冰箱保存。LB 固体培养基：每升 LB 培养基中加1.5 g 琼脂，高压灭菌器；冷却到 55 ℃后加入抗生素；倒入培养皿中。含有插入了外源基因的 pUC 19 质粒的 *E. coli* $DH_{5\alpha}$宿主细胞、IPTG 溶液（20% W/V）、X - gal 溶液（2% W/V）、含有合适抗生素的丰富肉汤琼脂平板、丰富肉汤顶层琼脂（表 6 - 1）。

表 6 - 1　丰富肉汤顶层琼脂平板

平板直径大小	肉汤琼脂添加量	X - gal	IPTG
90 mm	3 mL	40 μL	7 μL
150 mm	7 mL	100 μL	20 μL

【实验步骤】

（1）大肠杆菌在 LB 平板上划线，倒置平板于 37 ℃培养 1 ~ 2 天。

（2）从 LB 平板上挑取一单菌落接种到 2.5 mL LB 培养液中，37 ℃振荡培养过夜（约 250 转/分钟）。

（3）第二天，向装有 100 mL LB 培养基的 500 mL 锥形瓶中转接 2 mL 过夜细胞培养物。37 ℃振荡培养至 OD_{600}值达 0.5 至 0.6 之间（需 2 ~ 3 小时）。

（4）4 ℃，12 000 转/分钟离心 5 分钟收集细胞。用 1.2 mL 的 100 mmol/L 的冰冷 $CaCl_2$溶液重悬细胞冰浴 30 分钟。

（5）4 ℃，12 000 转/分钟离心 5 分钟收集细胞。

（6）用 250 μL 的 100 mmol/L 的冰冷 $CaCl_2$溶液重悬细胞。

（7）将试管置于冰上以待加入 DNA。

注释：感受态细胞对外界处理与高温特别敏感，因此操作要温和。

（8）为了对 $CaCl_2$处理过的细胞直接进行转化，用冷的微量移液枪枪头移取 200 μL 感受态细胞至无菌的冷的小聚丙烯试管中。向试管中加入 DNA（不超过 50 ng，体积不超过

10 μL）。轻微旋涡混合后冰上放置 30 分钟。

（9）将试管转至已预热到 42 ℃的循环水浴中的试管架上，精确放置 90 秒。不要摇动试管。

（10）将试管迅速转移至冰浴，让细胞冰却 1～2 分钟。

（11）向试管中加入 800 μL LB 培养基。37 ℃水浴培养 45 分钟以恢复和表达由质粒编码的抗生素性标记。

（12）将适量体积（每块 90 mm 平板不超过 100 μL）的转化细胞移至含有适当抗生素的 LB 琼脂培养基上。

（13）室温放置平板直至液体被吸收。

（14）倒置平板于 37 ℃培养 12～16 小时后应出现转化菌落。

（15）从培养箱中取出平板于 4 ℃储存几小时，使蓝颜色变深。

（16）对含有重组质粒的菌落进行鉴别。携有野生型质粒的菌落，具有 β－半乳糖苷酶活性，其中心为淡蓝色，外围为深蓝色。而携有重组质粒的菌落，无 β－半乳糖苷酶活性，呈乳白色或蛋壳蓝，有时中心会出现一个微弱的蓝点。挑选含有重组质粒的克隆并进行培养。

【实验结果】

重组体的筛选结果如图 6－1 所示。

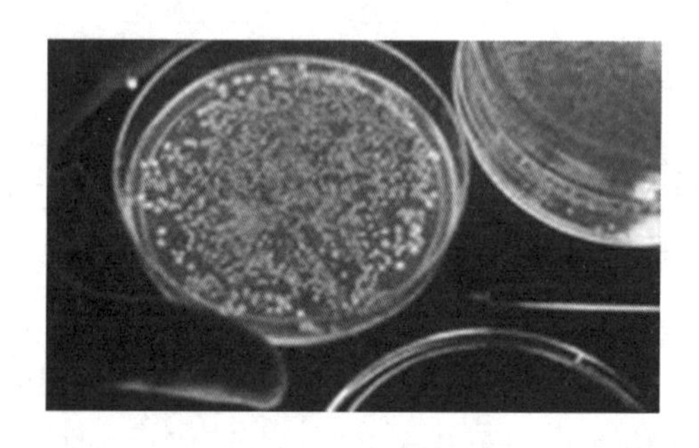

图 6－1　蓝白斑菌落筛选结果

【注意事项】

（1）获得高感受态细胞的关键是在全过程中使细胞保持冰冻。

（2）为获得高感受态，应当使用对数生长期的细胞。因此 OD_{600} 值不应高于 0.6。

（3）感受态大肠杆菌细胞很脆弱，处理时必须小心。

Ⅵ. Preparation of Competent *E. coli*, Transformation and Screening of Recombinant Cells

【Objectives】

(1) Learn how to prepare competent *E. coli* by using calcium chloride and transform it with the plasmid.

(2) Understand the principle and procedure of screening bacterial clone by α-Complementation.

【Principle】

Transformation results from the uptake of purified DNA by bacterial cells and is the most frequently used method for introducing recombinant DNA molecules into *E. coli*. However, DNA is normally unable to cross the membranes of most bacteria. In order to transfer DNA into cells through cell membrane, the bacterial cells must be in a competent state. Competence can be artificially induced in *E. coli* cells by treating them with calcium chloride prior to adding DNA. Ca^{2+} destabilizes cell membrane, and a calcium phosphate-DNA complex is formed, which adheres to the cell surface. DNA will be taken up during a heat-shock step when the cells are exposed briefly to a temperature of 42 ℃. It is believed that ice-cold calcium chloride treatment induces a "competence" state of bacteria and heat-shock opens the cell membrane to let DNA enter the cells.

This is a simple and quick method which yields up to 10^7 transformed colonies per microgram of supercoiled plasmid DNA. Many variations of this technique have been further developed to optimize the transformation efficiency. Transformation efficiency is the amount of cells transformed per microgram of DNA. In practice, much smaller amount of DNA are used (5 to 100 ng) since too much DNA inhibits the transformation process.

α-complementation occurs when two inactive fragments of β-galactosidase in *E. coli* associate to form a functional enzyme. Many plasmid vectors carry a short segment of DNA containing the coding information for the first 146 amino acids of β-galactosidase. Vectors of this type are used in host cells that express the carboxy-terminal portion of the β-galactosidase. Although neither the host nor the plasmid-encoded fragments of β-galactosidase are themselves active, they can associate to form an enzymatically active protein. Lac^+ bacteria that result from α-complementation can be easily recognized because they form blue colonies in the presence of the chromogenic substrate X-gal (5bromo-4-chloro-3-indoyl-β-D-galactoside). However, insertion of a fragment of foreign DNA into the polycloning site of plasmid almost inevitably results in production of an amino terminal fragment that is no longer capable of α-complementation. Bacteria carrying recombinant plasmids therefore cannot degrade X-gal and form white colonies. The development of this simple blue-white color test has greatly simplified the identification of recombinants constructed in plasmid vectors.

【Apparatus, materials and reagents】

1. Apparatus

Thermostat shaker, thermostat incubator, thermostat water bath, refrigerated centrifuge, laminar flow bench, −70 ℃ refrigerator and pipettes with the capacities of 10 μL, 100 μL and 1 000 μL respectively.

2. Materials

Calcium chloride ($CaCl_2$), tryptone, yeast extract, sodium chloride (NaCl), ampicillin, *E. coli* $DH_{5\alpha}$, plasmid pUC 19, sterile centrifuge tubes of 50 mL, tips, mini tubes, tubes, Petri dish and conical flask.

3. Reagents

0.1 mol/L $CaCl_2$, LB medium: dissolve 10 g tryptone, 0.5 g yeast extract and 0.5 g NaCl in 950 mL ddH_2O. Shake the container until the solute is completely dissolved, adjust the pH to 7.0 with NaOH (add 5 mol/L NaOH 200 μL per liter) and add deionized water to the total volume of 1 L and sterilize at 121 ℃ for 20 mins. Ampicillin (Amp) is prepared into 100 mg/mL solution with sterile water and stored at −20 ℃. LB plates with antibiotics: add 1.5 g agar per liter LB medium, autoclave; after cooling down to 55 ℃, add antibiotics; pour onto Petri dishes. *E. coli* $DH_{5\alpha}$ transformed with ligation mixture of insert to vector arms (pUC 19), IPTG solution (20% W/V), X-gal solution (2% W/V), broth agar plates containing appropriate amount of antibiotic and broth for top agar (Table Ⅵ−1).

Table Ⅵ−1 Components for top agar

Size of Plate Molten Top Agar	Amounts of Rich Broth Agar	X−gal	IPTG
90 mm	3 mL	40 μL	7 μL
150 mm	7 mL	100 μL	20 μL

【Procedures】

(1) Scribe *E. coli* cells onto LB plate, invert the plate and culture it at 37 ℃ for 1 – 2 days.

(2) Inoculate a single colony from an LB plate in a tube with 2.5 mL of LB medium. Incubate the tube overnight an 37 ℃ with shaking (approximately 250 rpm).

(3) On the following day, transfer 2 mL of the entire overnight culture into 100 mL LB medium in a 500 mL flask, culture the cells at 37 ℃ with shaking to OD_{600} between 0.4 and 0.6 (takes about 2 – 3 hours).

(4) Collect the cells by centrifugation at 12 000 rpm for 5 minutes at 4 ℃. Gently resuspend the cell pellets in 1.2 mL ice-cold 100 mmol/L calcium chloride. Incubate the resuspended cells on ice for 30 minutes.

(5) Collect the cells by centrifugation at 12 000 rpm for 5 minutes at 4 ℃.

(6) Gently resuspend the cell pellets in 250 μL ice-cold 100 mmol/L calcium chloride.

(7) Place the tubes on ice before adding DNA.

Note: treat the competent cells gently as they are highly sensitive to treatment and elevated temperature.

(8) To transform the calcium chloride-treated cells directly, transfer 200 μL of each suspension of competent cells to a sterile chilled polypropylene tube using a chilled micropipette tip. Add DNA sample to each tube (no more than 50 ng and size within 10 μL). Mix the contents of the tubes by swirling gently. Store the tubes on ice for 30 minutes.

(9) Transfer the tubes to a rack placed in a preheated 42 ℃ circulating water bath. Store the tubes in the rack for exactly 90 seconds. Do not shake the tubes.

(10) Rapidly transfer the tubes to the ice bath. Allow the cells to chill for 1 – 2 minutes.

(11) Add 800 μL of LB medium to each tube. Incubate the cultures for 45 minutes in a water bath set at 37 ℃ to allow the bacteria to recover and to express the antibiotic resistance marker encoded by the plasmid.

(12) Transfer appropriate amount (no more than 100 μL on each 90 mm plate) of transformed competent cells onto an LB medium agar containing the appropriate antibiotic.

(13) Store the plates at room temperature until the liquid has been absorbed.

(14) Invert the plates and incubate at 37 ℃. Transformed colonies should appear in 12 – 16 hours.

(15) Remove the plates from the incubator and store them for several hours at 4 ℃, to allow the blue color to develop.

(16) Identify colonies carrying recombinant plasmids. Colonies that carry wild-type plasmids contain active β-galactosidase. These colonies are pale blue in the center and dense blue at their periphery. Colonies that carry recombinant plasmids do not contain active β-galactosidase. These colonies are creamy-white or eggshell blue, sometimes with a faint blue spot in the center. Pick and culture colonies carrying recombinant plasmids.

【Result】

The result of recombinant screening is shown in Figure Ⅵ－1.

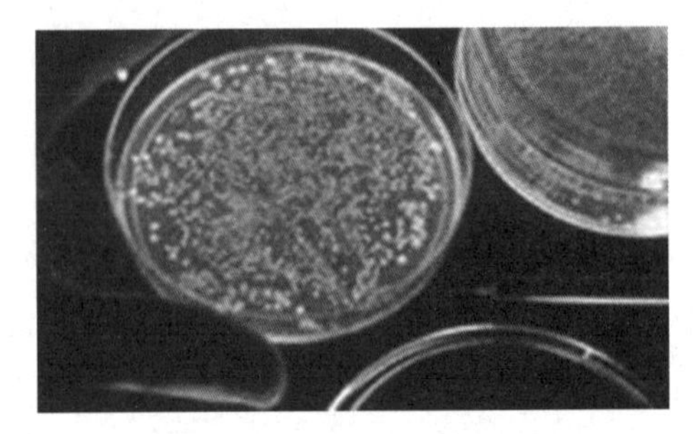

Figure Ⅵ－1　Result of blue-white screening

【Notes】

(1) It is critical to keep cells ice cold throughout this procedure to obtain high competency.

(2) To achieve high competency, cells in log phase should be used. Therefore, the OD_{600} should not be higher than 0.6.

(3) Competent *E. coli* cells are fragile and must be treated carefully.

七、基因组 DNA 的 Southern 杂交分析

【实验目的】

学习和掌握 Southern 转膜，随机引物法标记探针，用放射自显影方法取得分子杂交结果的方法。

【实验原理】

Southern 杂交是一项在复杂的背景基因中识别特异性 DNA 序列的重要技术之一。它是 Southern 于 1975 年首创的杂交方法。其基本原理是具有一定同源性的两条核酸单链 DNA（或 DNA 与 RNA）在一定的条件下可按碱基互补配对原则退火形成双链。杂交的双方是待测核酸序列及标记的探针。此杂交过程是高度特异性的。

首先将经限制性内切酶酶解、琼脂糖凝胶电泳分离的 DNA 片段在凝胶上经 NaOH 处理使之变性，然后用硝酸纤维素滤膜（nitrocellulose filter membrane，NC 膜）放在凝胶上，按原有顺序将条带转移至 NC 膜上并固定起来，这就是 Southern 转膜过程。

Southern 膜杂交是将吸附并固定在硝酸纤维素滤膜上的 DNA 片段与一个 ^{32}P 标记的 DNA 或 RNA 探针杂交，最后经过放射自显影从 X 光片上显现出杂交分子的区带。

本实验中，经琼脂糖凝胶电泳分离的 pUC 19 质粒 DNA，通过 Southern 转移，将其吸印在 NC 膜上，通过随机引物法，以同位素 ^{32}P 标记的 pUC 19 质粒 DNA 为探针，与 NC 膜上的质粒 DNA 进行杂交，再用放射自显影方法取得分子杂交的结果。

本实验所用待测核酸和探针都是 pUC 19 质粒 DNA，它们之间必然有同源性，可得到杂交带。这样可对初学者对 Southern 杂交方法进行训练。

【实验仪器、材料与试剂】

1. 仪器

恒温水浴器、烤箱、台式高速离心机、放射性污染监测器、琼脂糖凝胶电泳系统、高压灭菌锅、-70 ℃或 -20 ℃冰箱。

2. 材料

4 种 dNTP（其中一种为［a - ^{32}P］dCTP）、DNA 聚合酶 I 大片段（Klenow 酶）、随机

引物、氯化钠（NaCl）、柠檬酸钠、盐酸（HCl）、氢氧化钠（NaOH）、蔗糖、硼酸、溴酚蓝、三羟甲基氨基甲烷（Tris）、十二烷基硫酸钠（SDS）、磷酸二氢钠（NaH_2PO_4）、磷酸氢二钠（Na_2HPO_4）、显影粉、定影粉、溴化乙啶（EB）、微量加样器（20 μL，100 μL，1 000 μL 各一把）及吸头，小试管、Whatman No. 1 滤纸及普通滤纸、10 cm × 5 cm 玻璃板、硝酸纤维素滤膜（NC 膜）、卫生纸（或吸水纸）、裁纸刀、杂交盒、500 g 重物、X 光片夹及 X 光片、剪子、镊子、刀片、胶带、保鲜膜、直径 20 cm 的玻璃平皿、一次性手套、生化常用玻璃器皿、pUC 19 质粒 DNA。

3. 试剂

（1）20 × SSC（1 000 mL）：3 mol/L NaCl（175. 32 g），0. 3 mol/L 柠檬酸钠（88. 26 g），用 1 mol/L HCl 调至 pH 7. 0。

（2）变性溶液（500 mL）：含 1. 5 mol/L NaCl（43. 83 g）和 0. 5 mol/L NaOH（10 g）。

（3）中和溶液（500 mL）：0. 5 mol/L Tris（30. 27 g），3 mol/L NaCl（87. 66 g），用 1 mol/L HCl 调至 pH 7. 0。

（4）5 × TBE（250 mL）：13. 5 g Tris，6. 9 g 硼酸，0. 9 g EDTA – Na_2，用蒸馏水定容至 250 mL。

（5）TE（10 mL）：10 mmol/L Tris – HCl，1 mmol/L EDTA，pH 8. 0。

（6）0. 2 mol/L HCl（500 mL）。

（7）6 × 溴酚蓝载样液（1 mL）：0. 25% 溴酚蓝，40% 蔗糖。

（8）4 mol/L EDTA 1 mL。

（9）6 × SSC 杂交液贮液：20% SDS（50 mL），1 mol/L NaH_2PO_4（10 mL），1 mol/L Na_2HPO_4（50 mL），0. 5 mol/L EDTA（100 mL）。

（10）EB 染色液：0. 5 μg/mL 溴化乙啶。

【实验步骤】

1. 提取 pUC 19 质粒

2. 质粒 DNA 的琼脂糖凝胶电泳

3. Southern 转膜（以下操作要戴上一次性手套）

（1）用刀片切掉未用过的凝胶区域，并将凝胶切掉一个角（对点样顺序做个标记），然后转至玻璃平皿中。

（2）将胶置于 0. 2 mol/L HCl 中 10 分钟（加 HCl 的目的是将 DNA 部分脱嘌呤）。如 DNA <1 kb 可省却。缓缓摇动。当溴酚蓝由蓝色转变成橘黄色时停止，用蒸馏水漂洗 2 次。

（3）在室温下将凝胶浸泡在变性溶液中（目的是使 DNA 双链变性成单链），放置 40 分钟，使 DNA 充分变性并不断摇动。

（4）用蒸馏水漂洗 2 次。

（5）在室温下将凝胶转到另一玻璃平皿中，用中和溶液浸泡 40 分钟（目的是中和），不断摇动。

（6）用蒸馏水漂洗 2 次。

（7）将 2 个直径 20 cm 的玻璃平皿并排放置，里面倒上 20 × SSC，两个平皿上面放一块 10 × 15 cm^2 干净玻璃板，玻璃板上铺两张 Whatman No. 1 滤纸，滤纸与玻璃板同宽，滤纸的两个边垂入 20 × SSC 溶液中，使溶液不断地吸到滤纸上（注意：可用玻璃棒在滤纸上滚动以赶尽气泡，滤纸不能用手直接接触）。

（8）将 NC 膜裁成与凝胶大小一致，并在相应位置上剪掉一个角，然后用双蒸水浸湿，再转入 20 × SSC 溶液浸泡约半小时（注意：不要用手直接触摸 NC 膜）。

（9）将中和处理好的凝胶滑到已用 Whatman No. 1 滤纸铺好的玻璃板中央（用玻璃棒赶掉凝胶与滤纸间的气泡）。

（10）小心用镊子将 NC 膜准确放在凝胶上（此时 DNA 开始转移，不能再移动 NC 膜）。

（11）用玻璃棒仔细赶掉一切存在于 NC 膜与凝胶间的气泡。

（12）NC 膜上盖一张同样大小的普通滤纸，再次赶尽气泡。

（13）预先裁一沓同样大小的卫生纸，纸大小略比 NC 膜小 2 mm。压在滤纸上，约 10 cm厚。

（14）在卫生纸上放置一块玻璃板，玻璃板上放一个 500 g 的砝码或其他重物。

（15）让凝胶上的 DNA 转移 12 小时或过夜。

（16）将 NC 膜与凝胶剥离。弃去胶，把 NC 膜浸在 6 × SSC 溶液中，约 5 分钟后取出。

（17）把 NC 膜夹在 4 层普通滤纸中，置 65 ℃烘箱中烘烤 4 小时。

4. 探针标记

（1）取 20 ~ 100 μg DNA 加水至 45 μL，100 ℃变性 10 分钟，迅速插入冰中。

（2）以下操作均在同位素防护下进行：

准备一个微量离心管置于冰上，依次加入下列试剂（以下均为 Promega 公司提供，除 DNA 模板）：

标记 5 × buffer	10 μL
dATP · dGTP · dTTP 混合物	2 μL
变性的 DNA 模板	1.5 μL
不含核酸酶的 BSA	2 μL
[a − ^{32}P] dCTP（10 U/μL）	5 μL
Klenow 酶（5 U/μL）	1 μL
引物	2 μL
加无菌水至	50 μL

（3）用离心机轻轻混匀，然后在室温下标记 3 小时（放置同位素室）。

（4）终止：将完成标记后的反应物煮沸 2 分钟，立即放入冰浴中。加 5 μL 4 mol/L EDTA 后，冷藏（-20 ℃）或直接用于杂交。

5. Southern 杂交

（1）6×SSC 杂交液配制（1 000 mL）：取 43.8 g NaCl、36.9 g 柠檬酸钠，用 500 mL 水溶解，再添加以下溶液：

20% SDS	30 mL
1 mol/L NaH_2PO_4	8 mL
1 mol/L Na_2HPO_4	42 mL
Ficoll	1 g
PVP	1 g
BSA	1 g
0.5 mol/L EDTA	5 mL

定容至 1 000 mL，测一下 pH 值，如为 7.5 左右则正确。

SDS 能使蛋白质变性，减少蛋白质对 DNA 杂交的影响；NaH_2PO_4 与 Na_2HPO_4 提供 pH 7.5 缓冲条件；Ficoll、PVP 以及 BSA 的作用是防止单链 DNA 与硝酸纤维素膜的非特异性结合。

（2）鱼精 DNA 的变性：100 ℃水浴煮沸 10 分钟，迅速插入冰中。

（3）将已烘烤过的 NC 膜放入杂交盒中，加入预热的杂交液 10 mL，使杂交液刚好没过 NC 膜；同时加入变性鱼精 DNA（减少杂交时的背景干扰），使杂交液中的鱼精 DNA 的浓度为 0.5 mL/100 mL，65 ℃恒温水浴，预杂交 4～5 小时（同位素室进行，预杂交的目的是将非特异性序列封闭，从而使背景更清晰）。

（4）探针变性：完成标记后，加等体积 0.4 mol/L NaOH 混匀 10 分钟。

（5）将变性探针加入杂交液中，混匀（变性探针与膜上的特异性序列杂交）。

（6）65 ℃恒温水浴杂交 12 小时或过夜（同位素室进行）。

（7）洗膜（同位素室进行）。

一洗：2×SSC，0.1% SDS	10 分钟×2
二洗：1×SSC，0.1% SDS	10 分钟×2
三洗：0.5×SSC，0.1% SDS	10 分钟×2

（注意：洗膜的温度和时间、盐离子浓度可灵活掌握。洗膜的目的是将滤膜上未与 DNA 杂交的及非特异性杂交的探针分子从滤膜上洗去）

6. 放射自显影

（1）将滤膜用保鲜膜包好，置 X 光片夹中，并用同位素监测仪探测同位素强度，从而确定曝光时间。

（2）在暗室中将 NC 膜放在增感屏上（光面与 NC 膜接触），在滤膜上压上 X 光片，再压上增感屏后屏（光面与 X 光片接触）。为了防止滤膜和 X 光片移位，可在适当位置粘上胶带条。

（3）置 -70 ℃（或 -20 ℃）曝光，曝光时间由第 1 步决定。

（4）在暗室中冲片，显影→水洗→定影，用水洗净后吹干或晾干。

【实验结果】

将实验结果进行拍照处理。

【实验安排】

整个实验在 7 ~ 10 天内完成。

第一天：配试剂、提质粒。

第二天：电泳，Southern 转膜过夜。

第三天：取出 Southern 膜进行预杂交与杂交，并杂交过夜。

第四天：洗膜，放射自显影。在放射自显影过程中，可穿插安排别的实验。

Ⅶ. Southern Hybridization Analysis of Genomic DNA

【Objective】

Learn and master the method of Southern transfer membrane, random primer method to label the probe, and obtain the molecular hybridization results by radiographic autoradiography.

【Principle】

Southern hybridization is one of the most important techniques for identifying specific DNA sequences in complex background genes. It is a hybridization method pioneered by Southern in 1975. The basic principle is that two single-stranded DNA (or DNA and RNA) with certain homology can be annealed to form a double strand under certain conditions according to the principle of base complementarity. The two sides of the hybridization are the nucleic acid sequences to be tested and the labeled probes. This hybridization process is highly specific.

First, the DNA fragment, which has been digested by restriction endonuclease and separated by agarose gel electrophoresis, is denatured by NaOH treatment on the gel, and then a nitrocellulose filter membrane (NC membrane) is placed on the gel to transfer the bands to the NC membrane and fix them in the original order, which is Southern transfer process.

Southern membrane hybridization is the hybridization of a DNA fragment adsorbed and fixed on a nitrocellulose filter membrane with a ^{32}P-labeled DNA or RNA probe, followed by radiographic autoradiography to reveal the bands of the hybrid molecule on an X-ray film.

In this experiment, the pUC 19 plasmid DNA separated by agarose gel electrophoresis was blotted on the NC membrane by Southern transfer, and then hybridized with the plasmid DNA on the NC membrane by random priming method, using the isotope ^{32}P-labeled pUC 19 plasmid DNA as a probe, and then obtained the results of molecular hybridization by radiographic autoradiography.

In this experiment, the nucleic acid to be tested and the probe are both pUC 19 plasmid DNA, so there must be homology between them, and hybridization bands can be obtained. This can train the beginners on Southern hybridization method.

【Apparatus, materials and reagents】

1. Apparatus

Thermostatic water bath, oven, table top high-speed centrifuge, radioactive contamination monitor, agarose gel electrophoresis system, autoclave and −70 ℃ or −20 ℃ refrigerator.

2. Materials

4 types of dNTP (one of which is [a − ^{32}P] dCTP), the large fragment of DNA polymerase Ⅰ (Klenow enzyme), random primers, sodium chloride (NaCl), sodium citrate, hydrochloric acid (HCl), sodium hydroxide (NaOH), sucrose, boric acid, bromophenol blue, trimethylaminomethane (Tris), sodium dodecyl sulfate (SDS), sodium dihydrogen phosphate (NaH_2PO_4), disodium hydrogen phosphate (Na_2HPO_4), developing powder, fixing powder, ethidium bromide (EB), pipettes with capacities of 20 μL, 100 μL and 1 000 μL respectively, tips, mini-tubes, Whatman No. 1 filter paper and ordinary filter paper, 10 cm × 5 cm glass plates, nitrocellulose membrane (NC membrane), toilet paper (or absorbent paper), paper cutter, hybridization box, 500 g object, X-ray film clips, X-ray film, scissors, forceps, blades, tape, cling film, 20 cm diameter glass dish, disposable gloves, glassware commonly used in biochemistry, and pUC 19 plasmid DNA.

3. Reagents

(1) 20 × SSC (1 000 mL): Prepare the solution with 3 mol/L NaCl (175.32 g) and 0.3 mol/L sodium citrate (88.26 g). Adjust to pH 7.0 with 1 mol/L HCl.

(2) Denaturing solution (500 mL): Prepare the solution with 1.5 mol/L NaCl (43.83 g) and 0.5 mol/L NaOH (10 g).

(3) Neutralizing solution (500 mL): Prepare the solution with 0.5 mol/L Tris (30.27 g) and 3 mol/L NaCl (87.66 g). Adjust to pH 7.0 with 1 mol/L HCl.

(4) 5 × TBE (250 mL): Prepare the solution with 13.5 g Tris, 6.9 g boric acid and 0.9 g EDTA-Na_2 in ddH_2O. Dilute it to 250 mL in a volumetric flask.

(5) TE (10 mL): Prepare the solution with 10 mmol/L Tris-HCl and 1 mmol/L EDTA. Adjust pH to 8.0.

(6) 0.2 mol/L HCl (500 mL).

(7) 6 × Bromophenol blue carrier solution (1 mL): Prepare the solution with 0.25% bromophenol blue, 40% sucrose.

(8) 1 mL 4 mol/L EDTA.

(9) 6 × SSC Hybridization fluid storage solution: Prepare the solution with 20% SDS (50 mL), 1 mol/L NaH_2PO_4 (10 mL), 1 mol/L Na_2HPO_4 (50 mL) and 0.5 mol/L EDTA (100 mL).

(10) EB staining solution: 0.5 μg/mL ethidium bromide.

【Procedures】

1. Extract pUC 19 plasmid

2. Agarose gel electrophoresis of plasmid DNA

3. Southern transfer membrane (wear disposable gloves for the following operations)

(1) Cut off the unused gel area with a razor blade and cut off one corner of the gel (make a mark on the dot sample order). Transfer it to a glass flat dish.

(2) Place the gel in 0.2 mol/L HCl for 10 mins (the purpose of adding HCl is to partially depurate the DNA). If DNA <1 kb, it can be omitted. Slowly shake. Stop when the bromophenol blue turns from blue to orange and rinse 2 times with distilled water.

(3) Immerse the gel in denaturing solution at room temperature (the purpose is to denature DNA double-stranded into single-stranded) for 40 mins, allowing DNA to denature fully and shaking continuously.

(4) Rinse it 2 times with distilled water.

(5) Transfer the gel to another glass dish at room temperature and soak it in neutralization solution for 40 mins (the purpose is to neutralize), shaking continuously.

(6) Rinse it 2 times with distilled water.

(7) Place two 20 cm diameter glass dishes side by side, pour 20 × SSC inside; put a clean glass plate (10×15 cm^2) on top of the two dishes; lay two sheets of Whatman No. 1 filter paper on the glass plate; the width of the filter paper should be the same as the width of the glass plate, the two sides of the filter paper hanging into the 20 × SSC solution, so that the solution is constantly sucked onto the filter paper (Note: a glass rod can be used to drive out any air bubbles; the filter paper should not be directly touched by hand).

(8) Cut the NC membrane to the same size as the gel and cut off a corner at the corresponding position, then wet it with double distilled water and transfer it to 20 × SSC solution for about half an hour (Note: do not touch the NC membrane directly with your hands).

(9) Slide the neutralized gel onto the center of a glass plate that has been lined with Whatman No. 1 filter paper (use a glass rod to drive out any air bubbles between the gel and the filter paper).

(10) Carefully place the NC membrane exactly on the gel with forceps (at this point the DNA starts to transfer and the NC membrane should not be moved).

(11) Use a glass rod to carefully remove any air bubbles between the NC membrane and the gel.

(12) Cover the NC membrane with a piece of ordinary filter paper of the same size and again drive out any air bubbles.

(13) Pre-cut a stack of tissue paper of the same size, slightly smaller than the NC membrane by 2 mm, and press it onto the filter paper, about 10 cm thick.

(14) Place a glass plate on top of the tissue and a 500 g weight or other heavy object on the plate.

(15) Allow the DNA moleculars on the gel to transfer for 12 hours or overnight.

(16) Peel the NC membrane from the gel. Discard the gel and immerse the NC membrane in 6 × SSC solution for about 5 mins and then remove.

(17) Sandwich the NC membrane in 4 layers of plain filter paper and bake in an oven at 65 ℃ for 4 hours.

4. Probe labeling

(1) Add water to 20 – 100 μg of DNA to 45 μL in a mini tube. Denature it at 100 ℃ for 10 mins, and insert the tube rapidly into ice.

(2) The following operations are performed under isotope protection:

Prepare a microcentrifuge tube on ice and add the following reagents in order (all of the following are provided by Promega, except for the DNA template):

Labeled 5 × buffer	10 μL
dATP · dGTP · dTTP mixture	2 μL
Denatured DNA template	1.5 μL
BSA without nuclease	2 μL
[a – ^{32}P] dCTP (10 U/μL)	5 μL
Klenow enzymes (5 U/μL)	1 μL
Primers	2 μL
Add sterile water to	50 μL

(3) Mix gently with a centrifuge, then label the tube at room temperature for 3 hours (place in isotope chamber).

(4) Termination: Boil the reaction after completion of labeling for 2 mins and immediately put it into an ice bath. After adding 5 μL of 4 mol/L EDTA, refrigerate (–20 ℃) or use directly for hybridization.

5. Southern hybridization

(1) 6 × SSC hybridization solution preparation (1 000 mL): dissolve 43.8 g NaCl and 36.9 g sodium citrate in 500 mL of water. Add the following solutions:

20% SDS	30 mL
1 mol/L NaH_2PO_4	8 mL
1 mol/L Na_2HPO_4	42 mL
Ficoll	1 g
PVP	1 g
BSA	1 g
0. 5 mol/L EDTA	5 mL

Supplement ddH_2O to 1 000 mL. Test the pH value. If it is about 7. 5, it is correct.

The role of SDS is to denature the protein and reduce the effect of protein on DNA hybridization; NaH_2PO_4 and Na_2HPO_4 provide pH 7. 5 buffering conditions; the role of Ficoll, PVP and BSA is to prevent non-specific binding of single-stranded DNA to nitrocellulose membranes.

(2) Denaturation of fish sperm DNA: Boil the above solution for 10 mins at 100 ℃ in a water bath and insert it rapidly into ice.

(3) Put the baked NC membrane into the hybridization cassette. Add 10 mL of preheated hybridization solution to submerge the NC membrane just in the hybridization solution. Add denatured fish sperm DNA at the same time (to reduce background interference during hybridization). The concentration of fish sperm DNA in the hybridization solution is 0. 5 mL/100 mL of hybridization solution. Prehybridize it for 4 –5 hours in a constant temperature water bath at 65 ℃ (the isotope chamber is conducted, and the purpose of prehybridization is to close the non-specific sequence, thus making the background clearer).

(4) Denaturation of the probe: After completion of labeling, add an equal volume of 0. 4 mol/L NaOH and mix them for 10 mins.

(5) Add the denatured probe to the hybridization solution; mix them well (the denatured probe hybridizes with the specific sequence on the membrane).

(6) Hybridize in a constant temperature water bath at 65 ℃ for 12 hours or overnight (performed in the isotope chamber).

(7) Membrane washing (performed in the isotope chamber).

First wash: 2 ×SSC, 0. 1% SDS　10 mins ×2

Second wash: 1 ×SSC, 0. 1% SDS　10 mins ×2

Third wash: 0. 5 ×SSC, 0. 1% SDS　10 mins ×2

(Note: The temperature and time of membrane washing, salt ion concentration can be flexible; the purpose of membrane washing is to wash away the probe molecules from the membrane that are not hybridized with DNA and non-specific hybridization)

6. Radiographic autoradiography

(1) Wrap the membrane in cling film, place it in the X-ray film holder, and use an isotope monitor to detect the isotope intensity to determine the exposure time.

(2) In the darkroom, place the NC membrane on the intensifying screen (light surface in contact with the NC membrane), press the X-ray film on the filter film, and then press the rear screen of the intensifying screen (light surface in contact with the X-ray film). In order to prevent the filter film and X-ray film shift, stick the tape strip in the appropriate position.

(3) Expose the membrane at −70 ℃ (or −20 ℃). The exposure time depends on step 1.

(4) Develop the film in the darkroom, develop→ wash → fix, wash with water and blow dry or air dry.

【Result】

Take photos for experimental result.

【Arrangement】

The whole experiment needs to be completed between 7 – 10 days.

Day 1: Reagent preparation and plasmid extraction.

Day 2: Electrophoresis, Southern transfer membrane overnight.

Day 3: Remove Southern membrane for pre-hybridization and hybridization, and hybridize overnight.

Day 4: Wash the membrane and radiographic autoradiography. Other experiments can be arranged interspersed during the radiographic autoradiography.

八、外源基因在大肠杆菌中的诱导表达和检测

【实验目的】

（1）通过本实验了解外源基因在原核细胞中表达的特点和方法。

（2）学习 SDS－PAGE 的基本操作，学会用 SDS－PAGE 检测蛋白。

【实验原理】

将外源基因克隆在含有 *lac* 启动子的表达载体中，让其在大肠杆菌中表达。先让宿主菌生长，*lac* I 产生的阻遏蛋白与 *lac* 操纵基因结合，从而不能进行外源基因的转录及表达，此时宿主菌正常生长。然后向培养基中加入 *lac* 操纵子的诱导物 IPTG（异丙基硫代－β－D－半乳糖），阻遏蛋白不能与操纵基因结合，则 DNA 外源基因大量转录并高效表达。

表达蛋白可经 SDS－PAGE 检测。蛋白质与 SDS 结合后，均带有负电荷，在电场下，按相对分子质量大小在板状胶上排列。

【实验仪器、材料与试剂】

1. 仪器

恒温摇床、培养用锥形瓶、超净工作台、低温离心机、干热灭菌箱、普通恒压恒流电泳仪、平板电泳槽及配套的玻璃板、胶条、梳子。

2. 材料

克隆在大肠杆菌表达载体中的外源基因、酵母提取物、胰蛋白胨、氯化钠（NaCl）、甘油、氨苄西林（ampicillin，Amp）、异丙基硫代－β－D－半乳糖（IPTG）、十二烷基硫酸钠（SDS）、丙烯酰胺（Acr）、N，N’－亚甲基双丙烯酰胺（Bis）、三羟甲基氨基甲烷（Tris）、甘氨酸（Gly）、盐酸（HCl）、过硫酸铵（Aps）、四甲基乙二胺（TEMED）、低相对分子质量标准蛋白、预染标准蛋白（购自 Bio-lab 公司）、溴酚蓝、冰醋酸、乙醇、巯基乙醇、琼脂、考马斯亮蓝 R_{250}、滤菌膜、滤器、移液管、吸头、小试管。

3. 试剂

（1）LB（Luria-Bertani）培养基。

（2）TM 表达用培养基：

细菌培养用胰蛋白胨	12 g/L
细菌培养用酵母提取物	24 g/L
氯化钠	10 g/L
甘油	6 mL/L

用 Tris 调 pH 至 7.4，再用自来水补至 1 L，121 ℃高压蒸汽灭菌 20 分钟。

（3）50 mg/mL Amp 滤膜过滤于灭菌小试管中，－20 ℃贮存。

（4）1 mol/L IPTG 滤膜过滤于灭菌小试管中，－20 ℃贮存备用。

（5）1.5 mol/L Tris · HCl pH 8.8（4 ℃存放）。

（6）0.5 mol/L Tris · HCl pH 6.8（4 ℃存放）。

（7）10% SDS（室温存放）。

（8）30% Acr/Bis：29.2 g Acr 以及 0.8 g Bis，用双蒸水定容至 100 mL，过滤备用，4 ℃存放。

（9）10% Aps（－20 ℃存放）。

（10）2×上样缓冲液（室温存放）：

0.5 mol/L Tris · HCl pH 6.8	2 mL
甘油	2 mL
20%（W/V）SDS	2 mL
0.1 %溴酚蓝	0.5 mL
2－β－巯基乙醇	1.0 mL
双蒸水	2.5 mL

室温存放备用。

（11）5×电泳缓冲液（室温存放）：

Tris	7.5 g
Gly	36 g
SDS	2.5 g

双蒸水溶解，定容至 500 mL，使用时稀释 5 倍使用。

（12）染色液：0.2 g 考马斯亮蓝 R_{250} ＋84 mL 95% 乙醇 ＋20 mL 冰醋酸，定容至 200 mL，过滤备用。

（13）脱色液［医用酒精：冰醋酸：水＝4.5：0.5：5（V：V：V）］。

（14）保存液：7% 冰醋酸。

（15）封底胶：1% 琼脂（用蒸馏水配制）。

【实验步骤】

（1）含外源基因的表达菌株在 LB（含 50 μg/mL Amp）培养基中预培养过夜（注意

取菌株要在超净工作台上操作，一定要注意无菌）。

（2）100 mL TM 培养基加入 100 μL 50 mg/mL Amp，使终浓度达 50 μg/mL。按 1/100～1/50 比例加入过夜培养的上述 LB 培养液。于 37 ℃恒温摇床，250 转/分钟，培养 3 小时左右，使其 OD_{600} 值达 0.7～0.8（注意 OD_{600} 的值要视不同菌株中不同外源蛋白的表达情况而定，此 0.7～0.8 为本实验摸索出的经验值）。

（3）加入 50 μL 1 mol/L IPTG，终浓度达 0.5 mmol/L，进行外源基因的诱导表达。于 37 ℃恒温摇床，250 转/分钟，继续培养 4～5 小时（继续培养时间也需视蛋白表达情况而定）。

（4）4 ℃低温离心，4 000 转/分钟，20 分钟，收获菌体，弃上清液。菌体放 -20 ℃存放备用。

（5）配制分离胶。

①架好胶板，用 1.5 mm 胶条在两边隔开，用夹子固定，并用封底胶封底约 1 cm。

②按表 8-1 配制分离胶。

表 8-1　分离胶的成分

凝胶浓度/%	7.5	10	12	15	18	20
双蒸水/mL	9.6	8.1	6.7	4.7	2.7	1.5
1.5 mol/L Tris · HCl（pH 8.8）/mL	5	5	5	5	5	5
10%（W/V）SDS/μL	200	200	200	200	200	200
Acr/Bis（30%）/mL	5	6.65	8	10	12	13.2
TEMED/μL	10	10	10	10	10	10
10% Aps/μL	100	100	100	100	100	100
总体积/mL	20	20	20	20	20	20

混匀后加入两玻璃板夹缝中，并小心在胶面上加入 1 厘米蒸馏水，约 40 分钟，等胶自然凝聚后倾斜，倒出蒸馏水，并在两玻璃板夹缝中水平插入 1.5 毫米的梳子（在胶面上加入蒸馏水称水封，其目的是保持胶面平整和防止空气进入，影响凝胶）。

（6）按表 8-2 配制 4% 的浓缩胶。

表 8-2　4% 的浓缩胶的成分

双蒸水	6.1 mL
0.5 mol/L Tris · HCl pH 6.8	2.5 mL
10%（W/V）SDS	100 μL
Acr/Bis（30%）	1.3 mL

（续上表）

TEMED	10 μL
10% Aps	50 μL
总体积	10 mL

混匀后加入夹缝中，并没过梳子。待凝固后小心拔出梳子，用 100 μL 微量注射器抽取电极缓冲液冲洗梳子拔出后的加样凹槽底部，清除未凝的丙烯酰胺。

（7）样品制备。菌体样品与 2×上样缓冲液 1∶1 混匀，并在 100 ℃沸水浴中保温 3～5 分钟，取出待用。

（8）电泳。一孔加 10 μL 标准蛋白，一孔加样品（若做印迹，需加预染蛋白）。将玻璃板凝胶放入电泳槽中，并在槽中加入电极液，接通电源，电流调至 1 mA/孔，当样品进入分离胶时，调节电压使恒定在 120 V。当溴酚蓝移动到离底部约 0.5 cm 时，切断电源，停止电泳。将胶板从电泳槽中取出，小心地从玻璃板上取下胶。移去浓缩胶，将分离胶用考马斯亮蓝染色液染色，也可将此分离胶作印迹用。

（9）凝胶用染色液染色 2 小时，脱色过夜，换保存液保存胶。

【实验结果】

将 SDS－PAGE 检测表达蛋白结果拍照。

【实验安排】

第一天：（1）配制试剂和培养基，高压灭菌。

（2）预培养。

第二天：诱导表达。

第三天：SDS－PAGE 电泳。

Ⅷ. Inducible Expression and Detection of Exogenous Genes in *E. coli*

【Objectives】

(1) Learn the characteristics and methods of exogenous gene expression in prokaryotic cells through this experiment.

(2) Learn the basic operation of SDS-PAGE and learn to detect proteins by SDS-PAGE.

【Principle】

The exogenous gene is cloned in an expression vector containing the *lac* promoter and allowed to be expressed in *E. coli*. Firstly, the host bacterium is allowed to grow, and the blocker protein produced by *lac* Ⅰ is bound to the *lac* manipulator gene, so that the transcription and expression of the exogenous gene can not be carried out, and the host bacterium grows normally at this time. Then, IPTG (isopropylthio-β-D-galactose), the inducer of *lac* manipulator, is added to the medium, and the repressor protein can not bind to the manipulator gene, then the DNA exogenous gene is transcribed and expressed efficiently.

The expressed protein can be detected by SDS-PAGE. After binding to SDS, proteins are negatively charged and arranged on the plate gel according to their relative molecular masses under an electric field.

【Apparatus, materials and reagents】

1. Apparatus

Thermostatic shaker, conical flask for culture, ultra-clean worktable, cryogenic centrifuge, dry heat sterilizer, ordinary constant pressure and constant current electrophoresis system, plate electrophoresis tank and supporting glass plates, glue strips and combs.

2. Materials

Exogenous gene cloned in *E. coli* expression vector, yeast extract, tryptone, sodium chloride (NaCl), glycerol, ampicillin (Amp), isopropylthio-β-D-galactose (IPTG), sodium dodecyl sulfate (SDS), acrylamide (Acr), N, N'-methylenebisacrylamide (Bis), trimethylolaminomethane (Tris), glycine (Gly), hydrochloric acid (HCl), amine persulfate (Aps), tetramethylethylenediamine (TEMED), low relative molecular quality standard proteins, prestained standard proteins (purchased from Bio-lab), bromophenol blue, glacial acetic acid, ethanol, mercaptoethanol, agar, Coomassie Brilliant Blue R_{250}, bacterial filter membrane, filter, pipettes, pipette tips, mini tubes.

3. Reagents

(1) LB (Luria-Bertani) medium.

(2) Medium for TM expression:

Tryptone for bacterial culture	12 g/L
Yeast extract for bacterial culture	24 g/L
Sodium chloride	10 g/L
Glycerol	6 mL/L

Adjust pH to 7.4 with Tris, then make up to 1 L with water, and autoclave for 20 mins at 121 ℃.

(3) 50 mg/mL Amp filtered with filter membrane in sterilized mini tubes, stored at −20 ℃.

(4) 1 mol/L IPTG filtered with filter membrane in sterilized mini tubes and stored at −20 ℃.

(5) 1.5 mol/L Tris · HCl pH 8.8 (stored at 4 ℃).

(6) 0.5 mol/L Tris · HCl pH 6.8 (stored at 4 ℃).

(7) 10% SDS (stored at room temperature).

(8) 30% Acr/Bis: dissolve 29.2 g Acr and 0.8 g Bis in ddH_2O, dilute to 100 mL with double distilled water in a volumetric flask, filter and store it at 4 ℃.

(9) 10% Aps (stored at −20 ℃).

(10) Prepare 2 × sampling buffer (stored at room temperature) with the following solutions:

0.5 mol/L Tris · HCl pH 6.8	2 mL
Glycerol	2 mL
20% (W/V) SDS	2 mL
0.1% bromophenol blue	0.5 mL
2-β-mercaptoethanol	1.0 mL
Double-distilled water	2.5 mL

Store it at room temperature and reserve.

(11) 5 × Electrophoresis buffer (stored at room temperature):

Tris	7.5 g
Gly	36 g
SDS	2.5 g

Dissolve them in a beaker with double-distilled water and dilute to 500 mL in a volumetric flask. Dilute 5 times when using.

(12) Staining solution: prepare it with 0.2 g of Coomassie Brilliant Blue R_{250}, 84 mL of 95% ethanol and 20 mL of glacial acetic acid, dilute to 200 mL with ddH_2O. Filter it and reserve it for later use.

(13) Decolorization solution [Medical alcohol : glacial acetic acid : water = 4.5 : 0.5 : 5 (V : V : V)].

(14) Preservation solution: 7% glacial acetic acid.

(15) Seal gel: 1% agar (prepared with distilled water).

【Procedures】

(1) The expression strain containing the exogenous gene is pre-cultured overnight in LB (with 50 μg/mL Amp) medium (note that the strain is taken on an ultra-clean worktable bench and must be sterile).

(2) 100 mL of TM medium is added with 100 μL of 50 mg/mL Amp to make a final concentration of 50 μg/mL. Add overnight culture of the above LB culture at a ratio of 1/100 to 1/50. Incubate the culture at 37 ℃ in a thermostatic shaker at 250 rpm for 3 hours to reach the OD_{600} value of 0.7 – 0.8 (note that the value of OD_{600} depends on the expression of different exogenous proteins in different strains, and this value 0.7 – 0.8 is the empirical one for this experiment).

(3) 50 μL of 1 mol/L IPTG was added to a final concentration of 0.5 mmol/L for the induction of exogenous gene expression. Incubate at 37 ℃ in a thermostatic shaker at 250 rpm for 4 – 5 hours (the duration of incubation also depends on the protein expression).

(4) Centrifuge the above solution for 20 mins at 4 ℃ and 4 000 rpm. Harvest the bacteriophage and discard the supernatant. Store the bacteria at –20 ℃.

(5) Preparation of separation gel.

①Set up the gel board, separate it with 1.5 mm gel strips on both sides, fix it with clips and seal the bottom with sealing gel about 1 cm.

②Prepare the separation gel based on Table Ⅷ－1.

Table Ⅷ－1 The composition of separation gel

Gel concentration/%	7.5	10	12	15	18	20
Double-distilled water/mL	9.6	8.1	6.7	4.7	2.7	1.5
1.5 mol/L Tris · HCl (pH 8.8) /mL	5	5	5	5	5	5
10% (W/V) SDS/μL	200	200	200	200	200	200
Acr/Bis (30%) /mL	5	6.65	8	10	12	13.2
TEMED/μL	10	10	10	10	10	10
10% Aps/μL	100	100	100	100	100	100
Total volume/mL	20	20	20	20	20	20

After mixing, it should be added into the two glass joints and 1 cm of distilled water should be carefully added on the gel surface for about 40 mins, and the gel should be tilted and poured out of the distilled water after natural coalescence, and a 1.5 mm comb should be inserted horizontally into the joints of the two glass plates (the addition of distilled water on the glue surface is called water seal, the purpose of which is to keep the glue surface flat and prevent air from entering and affecting the gel).

(6) Prepare 4% concentrate gel based on Table Ⅷ－2.

Table Ⅷ－2 Composition of 4% concentrate gel

Double-distilled water	6.1 mL
0.5 mol/L Tris · HCl pH 6.8	2.5 mL
10% (W/V) SDS	100 μL
Acr/Bis (30%)	1.3 mL
TEMED	10 μL
10% Aps	50 μL
Total volume	10 mL

Mix them well and add the gel solution to the crevice, and immerse the comb. Then carefully pluck out the comb after solidification, and rinse the bottom of the addition groove after the comb is pulled out with 100 μL microsyringe to remove the uncoagulated acrylamide.

(7) Sample preparation. Bacterial samples are mixed with 2 × sampling buffer (1 : 1) and held in a boiling water bath at 100 ℃ for 3－5 mins. Take it out for use.

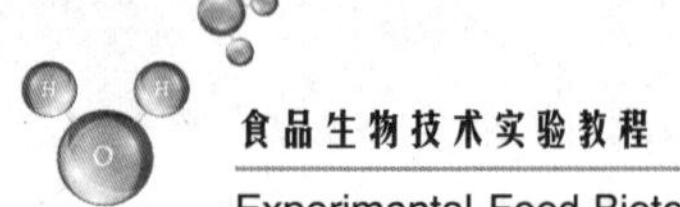

(8) Electrophoresis. Add 10 μL of standard protein to one well and sample to another (if blotting, add pre-stained protein). Put the glass plate with gel into the electrophoresis tank and add the electrode solution in the tank. Turn on the power and adjust the current to 1 mA/well. When the sample enters the separation gel, adjust the voltage to make it constant at 120 V. When the bromophenol blue moves to about 0.5 cm from the bottom, cut off the power and stop electrophoresis. Remove the plate with gel from the electrophoresis tank, and carefully remove the gel from the glass plate. Remove the concentrated gel and stain the separated gel with Coomassie Brilliant Blue staining solution, which can also be used for blotting.

(9) The gel is stained with the staining solution for 2 hours, decolorized overnight, and the gel is preserved by changing the preservation solution.

【Result】

Take the photo of the result of the expressed protein detected by SDS-PAGE.

【Arrangement】

Day 1: (1) Prepare the reagents and media. Autoclave them.

(2) Pre-culture.

Day 2: Induction of expression.

Day 3: SDS-PAGE electrophoresis.

九、溶菌酶的粗提取

【实验目的】

熟悉等电点沉淀法进行溶菌酶初级分离的操作方法和注意事项。

【实验原理】

蛋清中含水分87%，固形物13%；固形物中约90%是蛋白质，其中，卵白蛋白75%，卵类粘蛋白15%，卵粘蛋白7%，伴白蛋白3%。卵白蛋白、卵类粘蛋白和卵粘蛋白的等电点分别为4.7，3.9~4.5和3.9。溶菌酶少量存在于蛋清中，它溶于水，不溶于乙醚和丙酮，等电点为10.8，最适pH值为6.5。因此，可以通过调节pH值于卵白蛋白的等点电4.7，去除大部分的卵白蛋白，获得溶菌酶的粗提取液。

【实验仪器、材料与试剂】

1. 仪器

721型分光光度计、摇床、高速离心机。

2. 材料与试剂

200 mL烧杯、玻璃棒、漏斗、定性快速滤纸、200 mL量筒、50 mL离心管、鸡蛋、0.02 mol/L PBS（pH 7.0）、40%甘油。

【实验步骤】

（1）拿一个鸡蛋破壳取蛋清置于250 mL烧杯中，记录其体积V_1。

（2）加入1.5倍体积的pH 7.0 PBS缓冲液，搅拌均匀。取0.5 mL蛋清溶液并加入等量甘油于1 mL EP管，制备2管，-20 ℃备用（样品S_1）。

（3）样品S_1用冰醋酸调pH 4.7左右，充分搅拌，3 500转/分钟，离心20分钟，弃沉淀，转移上清至烧杯中。

（4）加入1倍体积的pH 7.0 PBS缓冲液，搅拌均匀，并用5 mol/L NaOH调pH 8.0，用滤纸过滤，取清液，测量并记录体积V_2。

（5）取0.5 mL清液并加入等量甘油于1 mL EP管中，制备2管，-20 ℃备用（样品S_2）。

（6）余下的清液放在100 mL烧杯中 –20 ℃冻存备用（样品 S_3）。

注：保存样品需明确标记名称、班级、组号、日期，最好使用标签纸。

【注意事项】

（1）等电点沉淀后利用离心的办法，要尽量除去沉淀。

（2）调节 pH 时要避免局部过酸。

（3）提取过程中尽量避免泡沫的产生。

【问题与思考】

（1）生物活性物质的粗提取有哪些方法？如何进行选择？

（2）本实验可否考虑其他粗提取方法？

（3）以粗分离中的离心和膜分离为例，说明其主要原理并举例。

（4）关于等电点沉淀的原理，pH 4.7 为何不是变性沉淀呢？

（5）为何先调 pH 4.7 左右，然后又需要将 pH 调回 8.0？

Ⅸ. Crude Extraction of Lysozyme

【Objective】

Be familiar with the operation and percautions of primary separation of lysozyme by isoelectric point precipitation method.

【Principle】

Albumen contains 87% of water and 13% of solids. The solids include about 90% of proteins, in which there are egg protein, ovomucoid, ovomucin and conalbumin account for 75%, 15%, 7%, and 3% respectively; the isoelectric points of egg protein, ovomucoid and ovomucin are respectively 4.7, 3.9 – 4.5 and 3.9. Lysozyme exists in egg white in small amount. This enzyme is soluble in water but insoluble in ester and acetone. Its isoelectric point is 10.8, and the preferable pH value is 6.5. Hence, it is possible to extract a crude solution of lysozyme by removing egg protein at its isoelectric point (4.7).

【Apparatus, materials and reagents】

1. Apparatus

Type 721 spectrophotometers, shakers, and high speed centrifuges.

2. Materials and reagents

200 mL beakers, glass rods, funnels, filtrate paper, 200 mL cylinders, 50 mL centrifuge tube, eggs, 0.02 mol/L PBS (pH 7.0), and 40% glycerol.

【Procedures】

(1) Break one egg, pour its albumen into a 250 mL beaker, and record its volume (V_1).

(2) Add 1.5 times of V_1 of PBS buffer (pH 7.0) and stir to homogeneous. Take 0.5 mL of albumen solution and the equal volume of glycerol into a 1 mL EP tube. Double it. The tubes are stored under –20 ℃ for further usage (Sample S_1).

(3) Sample S_1 is adjusted to pH 4.7 with glacial acetic acid and mixed well. Then it is cen-

trifuged at 3 500 rpm for 20 minutes. The precipitate is discarded and the supernatant is transferred to a new beaker.

(4) Add the equal volume of PBS buffer (pH 7.0) into the beaker and mix it well. Then use 5 mol/L NaOH solution to adjust the mixture solution to pH 8.0. Filtrate the mixture solution. Collect the filtrate and record its volume (V_2).

(5) Mix 0.5 mL of the filtrate with an equal volume of glycerol in a 1 mL EP tube. Double it too. Then store the tubes under −20 ℃ for further usage (Sample S_2).

(6) Place the left filtrate in a 100 mL beaker and store it under −20 ℃ for further usage (Sample S_3).

Note: Please mark names, classes, groups, date on the storing samples with labels.

【Notes】

(1) After precipitated at isoelectric point, the precipitate must be completely centrifuged and discarded.

(2) Avoid excess acid at local part when adjusting the pH value.

(3) Avoid production of foam during the procedure.

【Questions】

(1) Which method can be used in crude extraction of bioactive substances? How to select it?

(2) Are there any method applicable for crude extraction in this experiment?

(3) Take the isolation of centrifuge and membrane as examples to explain their principles and list some special examples.

(4) Why is not the precipitation at pH 4.7 a denaturing precipitation for albumen?

(5) Why do we adjust the pH value to 4.7 firstly, then back to 8.0?

十、溶菌酶分离纯化及酶活力测定

【实验目的】

（1）掌握离子层析的原理以及离子交换层析的操作方法。

（2）掌握离子交换树脂的再生和保存。

（3）掌握比色法测定溶菌酶的酶活。

【实验原理】

食物原料中的不同蛋白质分子之间的差异可能很微小，因此一些简单的纯化方法很难将一个酶蛋白完全纯化。一般来说，常根据酶蛋白的特性同时采用几种不同的方法结合进行纯化，如盐析、醇沉或等电点沉淀在纯化开始使用；而色谱分离比如离子交换色谱、凝胶色谱或者吸附色谱等一般在酶蛋白被某个沉淀法部分纯化后使用。特定的酶活或总酶活是酶蛋白纯化中的关键参数。酶的活力被定义成某个单位，比如每分钟产物形成的毫摩尔数（mmol/min）；而特定的酶活被定义为活力单位/mg 总蛋白。然而，目标酶的产量也很重要。例如，如果一个纯化程序使特定的酶活增加 100 倍，但只有 5% 酶活力单位被回收，而另一个程序可回收 90% 酶活力单位，但只使纯度增加 10 倍，则后者比前者有用。

本实验纯化的酶为溶菌酶。溶菌酶具有独特的电荷特性，其 pI 值高达 10.5，可以通过离子交换层析纯化。此外，溶菌酶是一种非常稳定的酶，即使在室温下保存几天也能保持催化活性。这种独特性加上蛋白质可轻易在当地杂货店或农场获得，使其成为介绍酶纯化技术的最佳选择。

柱色谱可以使蛋白质纯化数倍。阳离子交换树脂（CM – Sephadex）上有负电荷的羧酸基团，这种树脂能结合带正电荷的蛋白，如 pH 值中性条件下的溶菌酶。升高洗脱液的 pH，可以将溶菌酶从阳离子交换树脂上洗脱下来（为什么呢）。纯化时，将用到两种不同的阳离子交换树脂：①C – 25，蛋白质分子超过 30 kDa 的不能纯化；②C – 50，蛋白质分子超过 200 kDa 的不能纯化。其他色谱法如含惰性多孔微珠的凝胶过滤树脂（Sephadex G – 75），采用这种方法时，大于孔径的蛋白质分子将先直接流过柱子，而小于孔径的蛋白质分子将被滞留并且随后洗出。所有洗脱部分的溶菌酶活力、蛋白质含量将被监测以检验该纯化方法是否合适。

【实验仪器、材料与试剂】

1. 仪器

分部收集器、核酸蛋白质检测仪、记录仪、高速离心机（可用50 mL离心管）、冰箱、可见光分光光度计、摇床、烧杯、玻璃棒、漏斗、滤纸。

2. 材料及试剂

样品S_1（来自于实验九）、S_2（来自于实验九）、CM－Sepharose FF、0.02 mol/L pH 8.0的PBS、0.5 mol/L NaCl、烧杯（50 mL，250 mL）、枯草芽孢杆菌过夜培养物（37 ℃，18 h，160转/分钟，100或250 mL）、0.02 mol/L pH 8.0的PBS（测酶活用缓冲液，含50 mmol/L NaCl），0.02 mol/L pH 8.0的PBS（离子交换洗脱溶液，含0.5 mol/L NaCl）。

【实验步骤】

（一）溶菌酶的分离纯化

1. 装柱

取20 mL CM－Sepharose FF，1.0 cm×20 cm层析柱，以20 mmol/L PBS，pH 8.0为缓冲液装柱。注意不能使树脂露出水面，因为树脂露于空气中，当加入溶液时，树脂间隙中会产生气泡，而使交换不完全。对于重复使用的填料，需要先冲3倍柱床体积蒸馏水，将保存用的乙醇洗脱出来。

2. 平衡

用泵将3倍柱床体积的20 mmol/L PBS pH 8.0过柱。这一步的作用是使得柱床体系的内外达到平衡与均一，以利后续目的蛋白的结合。

3. 上样

用泵将S_2样品泵入层析柱，经过一段时间之后，蛋白质将通过离子交换的方式，与介质相互作用而挂柱。注意观察并记录280 nm下吸收值的变化。

4. 冲平

用3～4倍体积的0.02 mol/L pH 8.0的PBS洗涤离子交换柱至无穿透峰为止，流速约1.5 mL/min。在吸收峰下降段留样S_3，取约1 mL于EP管－20℃冻存备用。

5. 洗脱

用100 mL含0.5 mol/L NaCl的0.02 mol/L pH 8.0的PBS和100 mL 0.02 mol/L pH 8.0的PBS进行梯度洗脱。

收集洗脱液，记录洗脱时间和洗脱峰体积V_4。取0.5 mL洗脱液并加入等量40%甘油于1 mL EP管中，制备2管，－20 ℃备用（留样S_4）。

6. 再生

用泵将2倍柱体积的2 mol/L NaCl过柱，然后用4倍柱体积水洗。

注意：洗脱完成后，需要进行离子交换填料的再生处理。离子交换基团为一些盐所覆盖，影响下次的重复使用。因此，先用高浓度的盐将与介质结合紧密的杂质洗脱下来，然后再恢复其离子交换功能。对于 CM – Sepharose 而言，只需要用水过柱即可以使其离子交换功能得到恢复。

7. 保存

用泵将 2 倍柱体积的 20% 乙醇过柱。介质回收：用于蛋白质分离纯化的介质多保存于 20% 乙醇中。保存时需加入一些抑菌剂，如叠氮化钠等。对于本介质，采用 20% 乙醇保存即可。

对于洗脱下来的样品，我们无法确证是否含有溶菌酶，因此需要进行监测。一般采取活性监测的方式。本实验中利用溶菌酶水解枯草芽孢杆菌细胞壁，使得枯草芽孢杆菌裂解，以 595 nm 下光吸收值的变化来确定溶菌酶所在的吸收峰。

（二）溶菌酶的活性测定

取 6 mL 枯草芽孢杆菌的过夜培养物，3 500 转/分钟常温离心 5 分钟，弃上清，沉淀用6 mL 0.02 mol/L PBS 缓冲液悬浮，利用可见光分光光度计测其 OD_{595} 并记录（吸光度在 0.5 ~ 1.0 范围，如果读数过高可适当稀释）。比色皿编号和反应溶液的添加见表 10 – 1，其中第 1 支加入 PBS 作为仪器调零空白，第 2 支作为对照，3、4、5、6 分次加入溶菌酶样品（S_1 ~ S_4），每隔 30 秒测定一次在 595 nm 的吸光度值，3 分钟内至吸光度达到稳定。（因仪器数量有限，请各组错开时间进行，确保仪器准备完毕才加入酶液）在表 10 – 2 中记录它们的吸收值。

表 10 – 1　溶菌酶活力测定时添加的试剂

比色皿	1	2	3	4	5	6
PBS 缓冲液/mL	3	—	—	—	—	—
细胞 PBS 悬液/mL	—	3	2.9	2.9	2.9	2.9
分别用酶液 S_1 ~ S_4/mL	—	—	0.1	0.1	0.1	0.1

表 10 – 2　溶菌酶酶解过程中样品溶液的吸光度变化

时间/s	0	30	60	90	120	150	180
2 号							
3 号							
4 号							
5 号							
6 号							

观察悬液 OD_{595} 反应前后的变化。根据以下的酶活定义，测定溶菌酶 S_1 ~ S_4 活性。其中活力单位（U）：酶在室温（25 ℃），pH 8.0 条件下，OD_{595} 每分钟降低 0.001 为 1 个活力单位 U。处理数据后完成表 10 - 3：

表 10 - 3　样品的酶活

	S_1	S_2	S_3	S_4
酶活/（U · mL^{-1}）				

【注意事项】

（1）离子交换树脂在使用前需要再生，阴离子交换树脂以“碱酸碱”的顺序进行处理，阳离子交换树脂以“酸碱酸”的顺序进行处理和再生。装柱时要求粒度均匀，比较致密，柱床表面平整，柱中无裂缝、气泡和沟流的现象。

（2）加样蛋白浓度低于 20 mg/mL，上样体积小于柱体积的 1/3。

（3）在整个实验过程中，流速必须得到一定的控制。流速过大，会使填料压缩紧密，导致流速过低，层析柱有可能堵塞而实验失败；流速过小，实验时间过长，引起酶的活性变化。对于 CM - Sepharose FF 填料，最适流速在 100 cm/h 以下。因此，在这个实验中，控制流速为 2 mL/min。

（4）冲平过程一定不能省。在上样过程中，还有未挂柱的蛋白未能从色谱柱中完全流出，因此需先用缓冲液对整个色谱柱进行冲平，以便使未结合或结合不紧密的杂蛋白流出，以免干扰洗脱。

（5）洗脱时，梯度混合器中高盐溶液与低盐溶液的放置，依据洗脱目的进行。如果是离子交换，采用低盐上样、高盐洗脱的方式，就需要将低盐放置在靠近洗脱液出口的容器中。

【实验报告】

（1）如实完整地记录实验流程、现象及结果；分析记录仪上所绘制的峰型与所分离的蛋白质纯度的关系。

（2）实验报告中必须包含原始数据以及洗脱曲线图（同组可以复印同一张图）。

（3）本实验的数据处理：根据酶的动力学特性可知，在特定的时间段，酶促反应的速度是相等的，即产物浓度随时间呈线性变化。根据此特性，将所得数据在坐标纸上作图，以时间为横轴，OD_{595} 值为纵轴，将数据定位后，根据拟合直线的斜率，即为单位时间内 OD 值的下降值，然后再除以 0.001，即可得到所取测量酶液的活力单位。除以酶液体积，即可得到酶浓度（U/mL）。

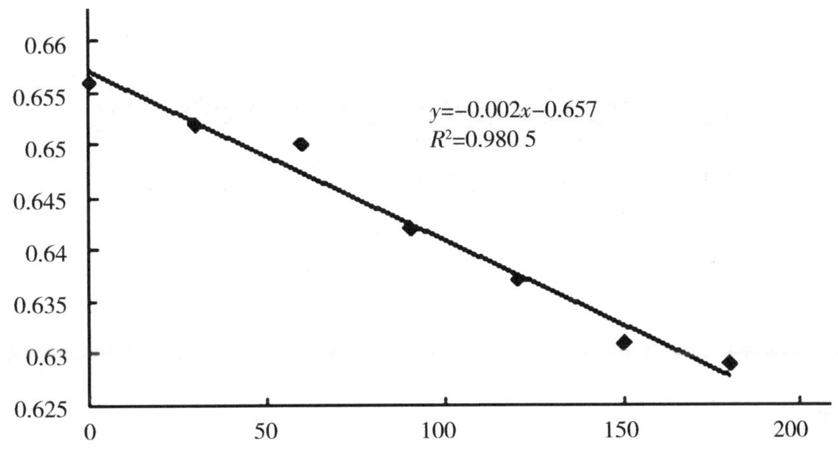

溶菌酶活力单位（U）= ΔOD/0. 001，其中 ΔOD 为一分钟内 OD 值的下降值。

【问题和思考】

（1）离子交换树脂填料如何再生？使用时的注意事项有哪些？

（2）为什么要有冲平这一步骤？

（3）为何要在整个分离纯化过程中不断监控酶的活性？

（4）分光光度法测定溶菌酶活力的原理是什么？注意事项有哪些？

X. Purification and Activity Determination of Lysozyme

【Objectives】

(1) Understand the principle and the operation of ion chromatography.

(2) Master the recovery and storage of ion exchange resin.

(3) Master the colorimetry method of determining the activity of lysozyme.

【Principle】

The difference between different protein molecules in food materials may be very small, so it is difficult to completely purify an enzyme protein by some simple purification methods. In general, different methods are used to purify simultaneously according to the characteristics of the enzyme protein, such as salting out, alcohol precipitation or isoelectric precipitation, which are used in the beginning of purification; and chromatographic separation, such as ion exchange chromatography, gel chromatography or adsorption chromatography, is usually used after partial purification of the enzyme protein by a certain precipitation method. Specific activity and total activity are the critical parameters in enzyme purification. Enzyme activity is defined as a unit, such as the number of millimoles of product generation per minute (mmol/min); while specific enzyme activity is defined as activity unit/mg of total protein. However, the yield of the desired enzyme is also important. For example, a purification step which yields a 100-fold increase in specific activity and a recovery of units (activity) of only 5% is not as useful as a step that yields a 10-fold purification with a recovery of 90%.

The enzyme purified in this experiment is lysozyme. The unique charge characteristics of lysozyme, which has an unusually high pI of 10.5, can be purified through ion exchange chromatography. Moreover, lysozyme is a remarkably stable enzyme and retains catalytic activity even after storage for several days at room temperature. These unique characteristics coupled with the fact that a major source of the protein can be cheaply obtained at the local grocery store or farm make it an excellent choice for an introduction to enzyme purification techniques.

The column chromatography should lead to a several-fold purification of the protein. The cation exchange resin (CM-Sephadex) has negatively charged carboxylic acid groups, which can bind proteins with positive charge, such as lysozyme at neutral pH. The lysozyme can be eluted by raising the pH (Why does a pH change elute the protein?) . Two different size beads will be used for cation exchange during the purification: C-25 with an exclusion limit of 30 kDa and C-50 with an exclusion limit of more than 200 kDa. The other column procedure involves a gel filtration resin (Sephadex G-75) consisting of inert porous beads. Proteins larger than the diameter of the pores will flow through the column whereas proteins smaller than the pores will be retained and thus eluted later. Eluted fractions for lysozyme activity, as well as total protein content will be monitored to test the purification procedure.

【Apparatus, materials and reagents】

1. Apparatus

Fraction collecter, nucleic acid and protein detector, recorder, high speed centrifuge (50 mL centrifuge tube applicable), refrigerator, visible spectrophotometer, shaker, beaker, glass rod, funnel, filter paper.

2. Materials and reagents

Samples (S_1 and S_2, from Experiment Ⅸ), CM-Sepharose FF, 0.02 mol/L PBS (pH 8.0), 0.5 mol/L NaCl, beaker (50 mL, 250 mL), overnight culture of *Bacillussubtilis* (37 ℃, 18 h, 160 rpm, 100/250 mL), 0.02 mol/L PBS (pH 8.0, containing 50 mmol/L NaCl), 0.02 mol/L PBS (pH 8.0, containing 0.5 mol/L NaCl) .

【Procedures】

i. Seperation, purification of lysozyme

1. Preparation of columns

Take 20 mL of CM-Sepharose FF to fill a column (1.0 cm × 20 cm) with 20 mmol/L PBS (pH 8.0) . Note that the resin cannot be exposed to the air, because if so, when the solution is added, there will be bubbles in the resin gap, which will make the exchange incomplete. For the recovery resin, 3 times column volumes of distilled water must be used to wash the column to make sure that ethanol, used to preserve resin, can be washed out.

2. Equilibrium

The 20 mmol/L PBS with 3 times the volume of column bed should be pumped through the column at pH 8.0. The function of this step is to make the column bed system inside and outside to achieve balance and uniformity, so as to facilitate the subsequent binding of the target protein.

3. Sample loading

Pump sample S_2 into the column. After a while, protein will interact with resin by ion exchange. Please observe and record the changes of absorbance at 280 nm.

4. Column washing

Wash the ion exchange column with 3 – 4 times the volume of 0.02 mol/L PBS (pH 8.0) until there is no breakthrough peak. The flow rate is about 1.5 mL/min. Take 1 mL of sample (S_3) in the falling section of absorption peak and freeze it in EP tube at –20 ℃ for later usage.

5. Elution

Use 100 mL of 0.02 mol/L PBS (pH 8.0) containing 0.5 mol/L NaCl and 100 mL of 0.02 mol/L PBS (pH 8.0) for gradient elution.

Collect the eluted solution by fraction collector, and record the eluting time and peak volume (V_4). Take 0.5 mL of eluted solution and an equal volume of 40% glycerol solution into a 1 mL EP tube and double it. Store two tubes (S_4) at –20 ℃ for later usage.

6. Recovery

Pump 2 times column volumes of 2 mol/L NaCl and then 4 times column volumes of distilled water to wash the column.

Note: after elution is done, the resin must be recovered since some salts will cover ion exchange groups and exert bad effect on the next usage. High concentration of salt solution can elute impurities from resin and then recover its function of ion exchange. As for CM-Sepharose, water can recover it.

7. Storage

Pump 2 times column volumes of 20% ethanol to wash the column. The recycle of resin: the resin used for purifying protein is usually stored in 20% ethanol containing some antibacterial agents, such as sodium azide. As for CM-Sepharose, 20% ethanol is enough for its storage.

We don't know which eluted fraction contains lysozyme. Hence, the enzymatic activity of each collected fraction must be determined. In this experiment, the activity of lysozyme is determined by its ability of breaking the cells of *Bacillus subtilis*. The absorption curve of lysozyme is described by absorbance at 595 nm.

ii. Determination of lysozyme activity

Take 6 mL of *Bacillus subtilis* overnight culture, and centrifuge at 3 500 rpm for 5 minutes at room temperature. Descant the supernatant. The precipitation is suspended with 6 mL of 0.02 mol/L PBS and its absorbance is determined with visible spectrophotometer at 595 nm (in the range of 0.5 to 1.0, dilute it if its absorbance is over high). Label cuvettes and add reaction solutions by following the instructions in Table Ⅹ－1. The cuvette only added PBS is labeled as No. 1 and control; the cuvette added cells in PBS as No. 2. Each of samples（$S_1 - S_4$）is added into cuvettes No. 3, 4, 5 and 6 in sequence for individual determination. The absorbance is determined every 30 s at 595 nm and it will be stable in 3 minutes. （Due to the limited number of instruments, please stagger the time to ensure that the instrument is ready before adding enzyme solution.）Record their absorbance in Table Ⅹ－2.

Table Ⅹ－1 Reaction solutions of determining lysozyme activity

Cuvette No.	1	2	3	4	5	6
PBS /mL	3	–	–	–	–	–
Cells in PBS /mL	–	3	2.9	2.9	2.9	2.9
Samples（$S_1 - S_4$）/mL	–	–	0.1	0.1	0.1	0.1

Table Ⅹ－2 The absorbance change of sample solution during lysozyme hydrolysis

Time/s	0	30	60	90	120	150	180
No. 2							
No. 3							
No. 4							
No. 5							
No. 6							

Observe the absorbance changes of OD_{595}. Determine the lysozyme（$S_1 - S_4$）activity by the following definition of lysozyme activity. One unit（U）of lysozyme activity is the amount that OD_{595} reducing 0.001 by lysozyme every minute at 25 ℃ and pH 8.0. Fill the calculated data in Table Ⅹ－3.

Table Ⅹ－3 Lysozyme activity of samples

	S_1	S_2	S_3	S_4
Lysozyme activity/（$U \cdot mL^{-1}$）				

【Notes】

（1）Before used, ion exchange resin needs to be recovered. An ion-exchange resin must be treated with alkali-acid-alkali in turn and cation-exchange resin treated with acid-alkali-acid in turn. When preparing a column, the size of resin must be unanimously compact without cracks, bubbles and grooves in column as well as the surface of column is flat.

(2) The concentration of loading protein must be less than 20 mg/mL and the volume of loading sample must be less than 1/3 column volume.

(3) During the experiment, the flow rate must be controlled. If it flows too fast, the resin will be heavily compacted, which will induce a very low flow rate. And more seriously, the column may be blocked up and the experiment will be failed, while the low flow rate makes the time of experiment become long and the activity of enzyme will lose. As for CM-Sepharose FF, the best flow rate is less than 100 cm/h. Therefore, in this experiment, the controlled flow rate is 2 mL/min.

(4) The step of washing column cannot be omitted since there are some proteins that have not combined with resin and do not totally flow out from the column. The step is to wash them out to remove their interruption.

(5) When the elution is in the process, the placement of high-salt solution and low-salt solution in gradient mixer is determined by the aim of elution. If ion exchange is performed, the low-salt solution will be placed near the outlet of elution solution since samples are loaded with low-salt solution.

【Experiment report】

(1) Please fully record the experiment procedure, phenomena and results. Analyze the relationship between the peaks that the recorder draws and the purity of separated proteins.

(2) The raw data and the eluted curve must be concluded in reports (the same group can share one copy).

(3) The treatment of data: the rate of enzymatic reaction is equal during a special period according to the characterization of enzymatic dynamics, which means the product concentration linearly varied with time. Therefore, the linear diagram can be drawn out with time as a horizontal axis and the values of OD_{595} as a vertical axis. After all data are located, the unit of enzymatic activity in solutions can be calculated by the slope of fitting line; the descending value of OD in a unit time divides 0.001. The concentration of enzyme in solutions can be obtained by the unit of enzymatic activity dividing the volume of enzyme solution (U/mL).

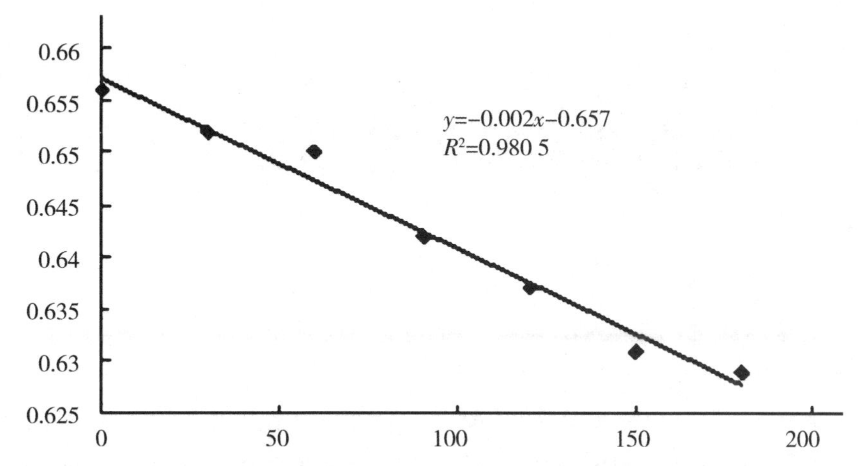

The unit of lysozyme activity (U) = ΔOD/0.001, in which, ΔOD means the descending value of OD in a minute.

【Questions】

(1) How to recover ion exchange resin? What should be noted in the recovery process?

(2) Why must the step of washing column be performed?

(3) Why must the lysozyme activity be monitored during the purification?

(4) What is the principle of determining lysozyme activity with spectrophotometer? What should be noted?

十一、溶菌酶纯度鉴定与分子质量测定

【实验目的】

（1）理解 SDS－聚丙烯酰胺凝胶（PAGE）测定蛋白质纯度与相对分子质量的原理。

（2）掌握电泳原理以及 SDS－PAGE 垂直板电泳分离蛋白质技术。

（3）熟悉凝胶染色与脱色方法，凝胶图谱的绘制与计算。

（4）了解在电泳中分子量标准物的使用。

【实验仪器、材料与试剂】

1．仪器

SDS－PAGE 电泳设备、电炉、脱色摇床。

2．材料

蛋清（S_1）、粗分离样品（S_2）、离子交换层析中穿透峰下降段样品（S_3）、离子交换层析洗脱峰样品（S_4）、橡胶手套。

3．试剂

2×上样缓冲液、10% SDS、30% 丙烯酰胺贮存液、分离胶缓冲液（1.5 mol/L Tris－HCl 缓冲液 pH 8.8）、浓缩胶缓冲液（0.5 mol/L Tris－HCl 缓冲液 pH 6.8）、10% 过硫酸铵（AP）、TEMED、Tris－Gly 电泳缓冲液 pH（8.3）、1% 琼脂糖溶液、染色液、脱色液。

【实验步骤】

1．蛋白样品的处理

取 200 μL 上述备用蛋白样品于带塞的小离心管中，加入 200 μL 的 2×上样缓冲液，混匀，轻轻盖上盖子，用封口膜封紧瓶口以免加热时迸出，在 100 ℃沸水浴中加热 3～5 分钟，取出冷却到室温后，10 000 转/分钟离心 10 分钟，取上清待用。

2．SDS－PAGE 实验步骤（适用于大板）

（1）把电泳玻璃板洗净，并在灌胶支架上固定好玻璃板，试样格（梳子）临用前用无水乙醇擦拭，让其挥发至干。

（2）用电泳缓冲液配制1%的琼脂糖凝胶，煮沸溶解后冷却至55 ℃左右，加入制板槽，槽内封板。

（3）按表11－1配好分离胶，用移液管快速加入电泳槽，大约5厘米，之后加少许蒸馏水，静置30分钟（注意：凝胶配制过程要迅速，催化剂TEMED要在注胶前加入，否则凝结无法注胶。注胶过程最好一次性完成，避免产生气泡。水封的目的是使分离胶上沿平直，并排除气泡。凝胶聚合好的标志是胶与水层之间形成清晰的界面）。

（4）倒出水并用滤纸把剩余的水分吸干，按表11－2配好浓缩胶，连续平稳加入浓缩胶至离槽边缘5毫米处，迅速插入样梳，静置25分钟（样梳需一次平稳插入，梳齿处不得有气泡，梳底需水平）。

（5）拔出样梳后，在上槽内加入Tris-Gly电泳缓冲液（pH 8.3），没过梳齿时可拆去底端的琼脂糖（要使锯齿孔内的气泡全部排出，否则会影响加样效果）。

（6）加样5个，空出第一个孔，从第二个开始按已编排好的顺序依次加入相应体积的样品和蛋白Marker。其中蛋清（S_1）和Marker加样量为2 μL，S_2、S_3和S_4加样量为10 μL。为了比较上样量的影响，可以加20 μL的S_1～S_4作为对照（注射器不可过低，以防刺破胶体；也不可过高，导致在样下沉时发生扩散。为避免边缘效应，最好选用中部的孔注样）。

（7）电泳槽中加入电泳缓冲液，接通电源，进行电泳。开始电压恒定在160 V，当样品进入分离胶后改为240 V，溴酚蓝距凝胶边缘约5毫米时，停止电泳。

（8）凝胶板剥离与染色：电泳结束后，用专用工具撬开短玻璃板，从凝胶板上切下一角作为加样标记，然后放在大培养皿内，加入染色液，42 ℃染色40分钟左右（剥胶时要小心，保持胶完好无损；凝胶染色要充分）。

（9）脱色：染色后的凝胶板用蒸馏水漂洗数次，再用脱色液脱色，直到蛋白质区带清晰。

（10）将凝胶室温保存于10%甘油水溶液中，至凝胶扫描分析。

（11）用凝胶扫描仪扫描并进行数据分析，求出溶菌酶的相对分子质量；分析各样品中溶菌酶相对含量的变化。

表11－1　分离胶的配制体系

分离胶（20 mL）	双蒸水	1.5 mol/L pH 8.8 Tris－HCl	胶贮液	10% SDS	TEMED	10% AP
12%	6.6 mL	5 mL	8 mL	200 μL	8 μL	200 μL

表11－2　浓缩胶的配制体系

浓缩胶（10 mL）	双蒸水	0.5 mol/L pH 6.8 Tris－HCl	胶贮液	10% SDS	TEMED	10% AP
4%	6.1 mL	2.5 mL	1.3 mL	100 μL	10 μL	100 μL

汇总所有实验结果，完成表 11－3。

表 11－3 实验结果

样品	体积/mL	蛋白浓度/ mg·mL^{-1}	总蛋白/mg	活力/ U·mL^{-1}	比活力/ U·mg^{-1}	总活力/U	回收率 /%	提纯倍数
S_1							100	1
S_2								
S_3								
S_4								

其中：总活力 = 比活力 × 体积。

$$回收率 = \frac{回收样品的酶活力}{总样品的酶活力} \times 100\%$$

$$提纯倍数 = \frac{提纯样品的比活力}{初始样品的比活力}$$

【注意事项】

（1）N，N’－亚甲双丙烯酰胺为神经毒性物质，可经皮肤直接吸收，使用时应避免其直接接触皮肤，必要时应戴手套，但其凝固后就变成无毒物质。

（2）安装电泳槽时要注意均匀用力旋紧固定螺丝，避免缓冲液渗漏。

（3）用琼脂（糖）封底及灌胶时不能有气泡，以免电泳时影响电流的通过。

（4）加样时样品不能超出凹形样品槽。加样槽中不能有气泡，如有气泡，可用注射器针头挑除。

【问题和思考】

（1）在不连续体系 SDS－PAGE 中，当分离胶加完后，需在其上加一层水，为什么？

（2）电泳缓冲液中甘氨酸的作用是什么？

（3）在不连续体系 SDS－PAGE 中，分离胶与浓缩胶中均含有 TEMED 和 AP，试述其作用。

（4）样品液为何在上样前需在沸水中加热几分钟？

Ⅺ. Determination of Purity and Molecular Weight of Lysozyme

【Objectives】

(1) To understand the principle of determining protein purity and molecular weight with SDS-PAGE.

(2) To master the principle of electrophoresis and the techniques of separating proteins with vertical electrophoresis by SDS-PAGE.

(3) To be familiar with the methods of dying and undying gel, the mapping and calculation of gel.

(4) To know about the application of standard molecular-weight substances.

【Apparatus, materials and reagents】

1. Apparatus

SDS-PAGE electrophoresis apparatus, electric furnace, shaker for discoloration.

2. Materials

Albumen (S_1), crude separated sample (S_2), the sample in the descending section of peak in ion exchange chromatography (S_3), the sample of eluted peak in ion exchange (S_4), rubber gloves.

3. Reagents

2 × loading buffer, 10% SDS, stock solution of 30% acrylamide, separating gel solution buffer (1.5 mol/L Tris-HCl, pH 8.8), stacking gel buffer (0.5 mol/L Tris-HCl, pH 6.8), 10% ammonium persulfate (AP), TEMED, electrophoretic Tris-Gly buffer (pH 8.3), 1% agarose solution, dying solution, undying solution.

【Procedures】

1. Preparation of protein sample

Pipette 200 μL of each sample into different tubes with stopper, then add 200 μL of 2 ×

loading buffer. Mix them and lightly put the stopper on. Tightly seal the tubes with film to avoid of spouting out in heating. Heat the tubes for 3 – 5 mins at boiling water bath and then take them out for cooling. After being cooled at normal temperature, they are centrifuged for 10 mins at 10 000 rpm. Take out the supernatant for further usage.

2. Steps of SDS-PAGE (applicable for big board)

(1) Wash the glass plate of electrophoresis and fix it in the bracket of pouring gel. Before using, the comb must be washed with absolute ethanol and then be dried by evaporation.

(2) The agarose gel is prepared with 1% agarose added into electrophoresis buffer. After dissolved by boiling, the gel is cooled to 55 ℃ and then poured into the vertical plate electrophoresis tank for sealing the tank.

(3) Prepare separating gel in proportion (Table Ⅺ – 1) and rapidly pipette it into the electrophoresis tank to a height of about 5 cm and then add some distilled water above. Make it stand for 30 mins (Note: gel preparation process must be quick and TEMED must be added before the gel poured in. Otherwise, TEMED will coagulate and block the gel poured in. The pouring-gel process should be completed without stop to avoid bubbles. The purpose of the water seal is to make the upper edge of the separating glue straight and eliminate bubbles. The good sign of gel polymerization is a clear interface between the glue and the water layer).

(4) Pour out water and use filter paper to absorb the left water. Prepare the stacking gel in proportion (Table Ⅺ – 2) . Continuously add the stacking gel into the tank to 5 mm from the edge, then quickly insert the comb and then stand for 25 mins (The comb should be inserted horizontally without stop, there should be no bubble at the combtooth, and the bottom should be horizontal).

(5) Pull the comb out and then pour the electrophoretic Tris-Gly buffer (pH 8.3) into the tank to submerge all combtooth holes. After that, remove the bottom agarose gel (make sure that all bubbles are removed from the combtooth holes to avoid some bad effects on loading samples) .

(6) Five samples and protein marker are individually loaded to the holes from the second start in order. The loading volume of S_1 and marker are 2 μL, and that of the others are 10 μL. If one wants to compare the effect of loading volume, 20 μL of S_1 to S_4 can be loaded as controls (The syringe should not be too low to prevent penetration of gel; it should not be too high to cause diffusion when the sample sinks. In order to avoid edge effect, it will be better to select the middle hole for injection) .

(7) Connect the power after add the electrophoretic buffer into the tank and start electrophoresis with 160 V at the beginning. After the samples enter the separating gel, adjust the power to 240 V. Stop the electrophoresis when bromophenol blue goes to about 5 mm from the edge of gel.

(8) Remove the gel plate and dye it: after the electrophoresis is over, open the short glass plate with a special tool. The gel plate should be carefully taken off and one piece of it should be cut off as a mark of loading samples. The gel should be then placed in a big culture dish and the dying solution should be added in for about 40 mins at 42 ℃ (pay attention when peeling the gel, keep it intact. The gel must be dyed adequately).

(9) Discoloration: wash the dyed gel for several times with distilled water, and then discolor with discoloring solution until the protein bands are clear.

(10) Keep the gel in a 10% glycerol solution at room temperature until the gel can be scanned and analyzed.

(11) Use a gel scanner to scan the gel and analyze its data to calculate the molecular weight of lysozyme. Meanwhile, analyze the changes of relative quantity of lysozyme in each sample.

Table XI-1 Composition of separating gel

Separating gel (20 mL)	Double distilled water	1.5 mol/L pH 8.8 Tris-HCl	Stock solution for gel	10% SDS	TEMED	10% AP
12%	6.6 mL	5 mL	8 mL	200 μL	8 μL	200 μL

Table XI-2 Composition of stacking gel

Stacking gel (10 mL)	Double distilled water	0.5 mol/L pH 6.8 Tris-HCl	Stock solution for gel	10% SDS	TEMED	10% AP
4%	6.1 mL	2.5 mL	1.3 mL	100 μL	10 μL	100 μL

Fill the data in the following table (Table XI-3).

Table XI-3 Experiment results

Samples	Volume /mL	Concentration of protein /mg·mL^{-1}	Total protein /mg	Activity /U·mL^{-1}	Specific activity /U·mg^{-1}	Total activity /U	Recovery rate/%	Time of purity
S_1							100	1
S_2								
S_3								
S_4								

In Table Ⅺ－3, total activity = specific activity × the volume of sample.

Recovery rate = activity of recovery sample / total activity ×100%.

Time of purity = specific activity after a sample purified / specific activity before a sample purified.

【Notes】

(1) N, N' -methylenebisacrylamide is a neurotoxin and can be absorbed through skin. When using it, take gloves to prevent it contacting the skin. But it won't be toxic after it becomes coagulated.

(2) To avoid leaking the buffer, pay attention to fixing screw when setting up the electrophoresis tank.

(3) Bubbles are not allowed when using agarose to seal the bottom of tank for avoiding a bad effect on electric current.

(4) Samples should not excess the sample holes when loaded. Bubbles can't be allowed either in the loading holes. If there are bubbles, use syringe needle to remove them.

【Questions】

(1) Why water is overlapped on the separating gel in the discontinuous system of SDS-PAGE?

(2) What's the function of Gly in electrophoresis buffer?

(3) What are the functions of TEMED and AP in the separating gel and stacking gel in discontinuous system of SDS－PAGE?

(4) Why must the sample solutions be heated in a boiling bath before they are loaded?

十二、固定化淀粉酶与活力测定

【实验目的】

学会用交联法制备固定化酶的操作技术。

【实验原理】

制备固定化酶的方法很多，利用双功能试剂或多功能试剂在酶分子间，酶分子与惰性蛋白间，或酶分子与载体间进行交联反应，以共价键制备固定化酶的方法称为交联法，本实验即采用这种方法。交联剂为戊二醇，载体为甲壳素。

【实验仪器、材料与试剂】

1. 仪器

恒温水浴锅、恒温振摇仪。

2. 材料

烧杯、容量瓶、试管。

3. 试剂

（1）5%戊二醇和甲壳素。

（2）碘原液：称取碘 1.1 g，碘化钾 2.2 g，置于小烧杯中，加 10 mL 蒸馏水使之溶解，然后转入容量瓶中。再加少量的蒸馏水洗涤烧杯数次，洗涤液均转入容量瓶中，最后定容至 50 mL。摇匀后放于棕色试管中备用。

（3）比色碘溶液：取碘原液 2 mL，加碘化钾 20 g，再用蒸馏水定容至 5 000 mL。

（4）2%淀粉溶液：称取 2 g 可溶性淀粉，放入小烧杯中，加少量蒸馏水做成悬浮液。然后在搅拌下注入沸腾的蒸馏水中，继续煮沸 1 分钟，冷却后加蒸馏水定容至 1 000 mL。

（5）pH 6.0 磷酸氢二钠—柠檬酸缓冲液：称取磷酸氢二钠（$Na_2HPO_4 \cdot 12H_2O$）45.23 g，柠檬酸（$C_6H_8O_7 \cdot H_2O$）8.07 g，先在烧杯中使之溶解，然后转入容量瓶中定容至1 000 mL。

（6）标准终点色溶液。

A 液：精确称取氯化钴（$CoCl_2 \cdot 6H_2O$）40.249 3 g 和重铬酸钾（$K_2Cr_2O_7$）0.487 8 g，用

蒸馏水定容至 500 mL。

B 液：精确称取络黑 T 40 mg，用蒸馏水定容至 100 mL。

同时取 A 液 40 mL、B 液 5 mL，混合后置于冰箱中待用。混合液在 15 天内使用有效。

【实验步骤】

1．酶液的制备

精确称取 α－淀粉酶 2g，先用少量 40 ℃ pH 6.0 的磷酸氢二钠—柠檬酸缓冲液溶解，溶解过程中轻轻用玻璃棒捣研。然后将上层液小心倾入 100 mL 容量瓶，沉渣部分再加入少量上述缓冲液，如此反复捣研 3～4 次。最后，将溶液与残渣全部移入容量瓶中，用缓冲液先定容摇匀后，通过四层纱布过滤，溶液供测定使用。

2．固定化酶的制备

（1）称取 50 mg 粉末甲壳素，加入 5% 戊二醛 10 mL，调节 pH 为 8.5，搅拌均匀后，于 25 ℃恒温振摇 1 小时。取出后，倾去戊二醛，然后以蒸馏水洗涤，倾去清液，以除去多余的交联剂。

（2）取前面制备的酶液 10 mL，与上述处理的甲壳素混合均匀，25 ℃恒温振摇 1 小时，然后 4 ℃冰箱放置过夜。

（3）取出后，4 000 转/分钟离心分离，倾去清液，蒸馏水洗涤，可得固定化酶。

3．固定化 α－淀粉酶活力的测定及活力回收率的计算

（1）首先用吸管取 1 mL 的标准终点色溶液，加至白瓷板的空穴内，作为终点参照的标准。

（2）固定前总酶活力测定：取 20 mL 2% 的可溶淀粉液与 5 mL pH 6.0 的磷酸氢二钠—柠檬酸缓冲液，加入一支大试管中。将试管置于 60 ℃水浴 5 分钟。然后加入前面制备的酶液 0.5 mL。摇匀后，立即用秒表记录时间。此后，每经一段时间，用吸管吸出 0.2 mL 反应液，加入预先盛入稀碘液的白瓷板中。当穴内颜色反应由紫色逐渐变为红棕色并与标准色相同时，即为反应终点，记录反应到达终点的时间。

（3）固定化酶活力测定：取 20 mL 2% 的可溶淀粉液与 5 mL pH 6.0 的磷酸氢二钠—柠檬酸缓冲液，加入一支大试管中。将试管置于 60 ℃水浴 5 分钟。然后加入前面制备的固定化酶。摇匀后，立即用秒表记录时间。此后，不断振摇，每经一段时间，用吸管吸出 0.2 mL反应液，加入预先盛入稀碘液的白瓷板中。当穴内颜色反应由紫色逐渐变为红棕色并与标准色相同时，即为反应终点，记录反应到达终点的时间。

【注意事项】

测定酶制剂时，应先在40 ℃水浴中悬浮2小时，再用纱布过滤。每次操作所用的纱布数要一致。

【实验结果】

（1）酶活力计算：以60 ℃、pH 6.0的条件下，每小时水解1 g淀粉的酶量为一个活力单位。

固定前原酶活力 $=60/T\times20\times2\%\times n/0.5$

固定后的酶活力 $=60/T\times20\times2\%\times n/10$

T：反应到终点时的时间（分）

n：酶粉稀释的倍数

（2）固定化后酶活力回收率计算：

酶活力回收率 =（固定后的酶活力单位/固定前原酶活力单位）×100%

XII. Immobilization and Activity Determination of Amylase

【Objective】

Learn the technique of cross-linking to prepare immobilized enzyme.

【Principle】

There are many methods for preparing immobilized enzymes, using bifunctional reagents or multifunctional reagents to conduct cross-linking reactions between enzyme molecules, between enzyme molecules and inert proteins, or between enzyme molecules and carriers, and prepare immobilized enzymes by covalent bonds. This experiment adopts the cross-linking method. The cross-linking agent is pentanediol, and the carrier is chitin.

【Apparatus, materials and reagents】

1. Apparatus

Thermostatic water bath and thermostatic shaker.

2. Materials

Beaker, volumetric flask and test tube.

3. Reagents

(1) 5% pentanediol and chitin.

(2) Iodine stock solution: weigh 1.1 g of iodine and 2.2 g of potassium iodide, place them in a small beaker, add 10 mL of distilled water to dissolve them, and then transfer them to a volumetric flask. Add a small amount of distilled water to wash the beaker several times, transfer the washing liquid to the volumetric flask, and finally dilute the volume to 50 mL. Shake it well and place it in a brown test tube for later use.

(3) Colorimetric iodine solution: take 2 mL of iodine stock solution, add 20 g of potassium iodide, and dilute the volume to 5 000 mL with distilled water.

(4) 2% starch solution: weigh 2 g of soluble starch, put it into a small beaker, and add a

small amount of distilled water to make a suspension. Then pour it into boiling distilled water under stirring, continue to boil for 1 minute, and add distilled water to make the volume up to 1 000 mL after cooling.

(5) Disodium hydrogen phosphate-citric acid (pH 6.0): weigh 45.23 g of disodium hydrogen phosphate ($Na_2HPO_4 \cdot 12H_2O$) and 8.07 g of citric acid ($C_6H_8O_7 \cdot H_2O$), dissolve it in a beaker, and then transfer it to a volumetric flask to measure the volume to 1 000 mL.

(6) Standard end point color solution.

Solution A: weigh accurately 40.249 3 g of cobalt chloride ($CoCl_2 \cdot 6H_2O$) and 0.487 8 g of potassium dichromate ($K_2Cr_2O_7$), and dilute the volume to 500 mL with distilled water.

Solution B: weigh 40 mg of Luohei T accurately, dissolve them and dilute the volume to 100 mL with distilled water.

At the same time, take 40 mL of solution A and 5 mL of solution B, mix them and place them in the refrigerator for later use. The mixture is effective within 15 days.

【Procedures】

1. Preparation of enzyme solution

Accurately weigh 2 g of α-amylase, first dissolve it with a small amount of 40 ℃ disodium hydrogen phosphate-citric acid buffer (pH 6.0), and then gently pound it with a glass rod during the dissolution process. Pour the upper layer carefully into a 100 mL volumetric flask, add a small amount of the above mentioned buffer to the sediment part, and repeat the pounding 3 – 4 times. Finally, transfer all the solution and residues into a volumetric flask, and shake the flask with a buffer solution, then filter through 4 layers of gauze, and the solution is to be used for measurement.

2. Preparation of immobilized enzyme

(1) Weigh 50 mg of powdered chitin, add 10 mL of 5% glutaraldehyde, adjust the pH to 8.5, stir well, and shake at a constant temperature of 25 ℃ for 1 hour. After taking it out, decant glutaraldehyde, and then wash with distilled water, decant the clear liquid to remove the excess cross-linking agent.

(2) Take 10 mL of the previously prepared enzyme solution, mix it with the chitin treated above, and shake it at a constant temperature of 25 ℃ for 1 hour, and then place it in a refrigerator at 4 ℃ overnight.

(3) After taking it out, centrifuge at 4 000 rpm, pour off the clear liquid, and wash with distilled water to obtain the immobilized enzyme.

3. Determination of activity of immobilized α-amylase and calculation of activity recovery rate

(1) First use a pipette to take 1 mL of the standard end point color solution and add it to the cavity of the white porcelain plate as the end point reference standard.

(2) Determination of total enzyme activity before immobilization: take 20 mL of 2 % soluble starch solution and 5 mL of disodium hydrogen phosphate-citric acid buffer (pH 6.0), and add them to a large test tube. Place the test tube in a 60 ℃ water bath for 5 minutes. Then add 0.5 mL of the previously prepared enzyme solution. After shaking well, immediately record the time with a stopwatch. Thereafter, every certain period of time, use a straw to suck out 0.2 mL of the reaction solution and add it to a white porcelain plate preliminarily filled with dilute iodine solution. When the color reaction in the hole gradually changes from purple to reddish brown and finally becomes the same as the standard color, it is the end of the reaction. Record the time when the reaction reaches the end.

(3) Determination of immobilized enzyme activity: take 20 mL of 2% soluble starch solution and 5 mL of disodium hydrogen phosphate-citric acid buffer (pH 6.0), and add them to a large test tube. Place the test tube in a 60 ℃ water bath for 5 minutes. Then add the previously prepared immobilized enzyme. Shake it well and immediately record the time with a stopwatch. After that, shake continuously, and every certain period of time, use a straw to suck out 0.2 mL of the reaction solution and add it to the white porcelain plate preliminarily filled with dilute iodine solution. When the color reaction in the hole gradually changes from purple to reddish brown and finally becomes the same as the standard color, it is the end of the reaction. Record the time when the reaction reaches the end.

【Notes】

When determining the enzyme preparation, it should be suspended in a water bath at 40 ℃ for 2 hours, and then filtered with gauze. The number of gauze used in each operation must be the same.

【Results】

(1) Enzyme activity calculation: under the conditions of 60 ℃ and pH 6.0, the amount of enzyme that hydrolyzes 1 g of starch per hour is one unit of activity.

Unit of enzyme activity before immobilization $= 60/T \times 20 \times 2\% \times n/0.5$

Unit of enzyme activity after immobilization $= 60/T \times 20 \times 2\% \times n/10$

T: time to end the reaction (minutes)

n: Dilution ratio of enzyme

(2) Calculation recovery rate of enzyme activity after immobilization:

Recovery rate of enzyme activity = (enzyme activity after immobilization / original enzyme activity before immobilization) $\times 100\%$.

十三、大肠杆菌中非天然氨基酸的整体掺入

【实验目的】

（1）理解大肠杆菌中掺入非天然氨基酸（色氨酸类似物）的原理。

（2）熟悉蛋白质的表达和纯化实验技术。

（3）了解非天然氨基酸掺入蛋白质的程度的分析方法。

【实验原理】

色氨酸类似物可直接掺入大肠杆菌中。因为色氨酰转移RNA合成酶没有编辑结构域，天然氨基酸与类似物的区分只是在结构上。枯草杆菌色氨酰转移RNA合成酶可以相对更有效地结合（装载）荧光化的色氨酸类似物；4－氟色氨酸（4fW）的结合比色氨酸低6倍，而5－氟色氨酸（5fW）的结合比色氨酸低74倍。与此类似，已经知道色氨酸类似物能够进入细胞，并在类似物的毒性起作用之前支持最少几代的生长。为了使色氨酸类似物有效掺入，需要使用带有防止生物合成的突变细菌菌株。

【实验仪器、材料与试剂】

1. 仪器

（1）Microcon浓缩器，拦截的相对分子质量为10 000。

（2）Ni－次氮基三乙酸（Ni－NTA）树脂和蛋白质纯化柱。

（3）Centri－Sep尺寸排斥柱。

（4）用于DNA的琼脂糖凝胶电泳设备和十二烷基硫酸钠（SDS）凝胶电泳（PAGE）设备。

（5）HPLC、HPLC－电喷电离（ESI）和质谱设备。

（6）酶标仪（一般称微板阅读器）。

2. 材料

（1）大肠杆菌菌株C600 Δ*trp*E（*thi*－1 *thr*－1 *leu*B6 *lac*Y1 *ton*A21 *sup*E44 *mcr*A Δ*trp*E）和衍生菌株C600p（C600 Δ*trp*E + pUC 18）、C600pGSR（C600 Δ*trp*E + pG － SR）、C600F（C600 Δ*trp*E F’ KanR）和C600F（DE3）（C600FλDE3溶源体）。用于转化的菌株

有 $DH_{5\alpha}F'$ 和 TOP10。

（2）Luria – Bertani 培养基（每升含：10 g 胰蛋白胨、5 g 酵母抽提物、10 g 氯化钠和 1.5% 菌用琼脂糖用于培养板）和基本培养基 M9（5 × 原液，每升含：30 g 磷酸氢二钠、15 g 磷酸二氢钾、5 g 氯化铵、2.5 g 氯化钠和 1.5% 菌用琼脂糖用于培养板），补充 20 μg/mL氨基酸和 0.000 5% 维生素 B_1。富集培养基和基本培养基都如标示的那样补充抗生素：50 μg/mL 氨苄西林（Amp）或卡那霉素（Kan）。

（3）色氨酸类似物：4 – 氟色氨酸（4fW）。

（4）用于蛋白质高表达的质粒：pGEX – KG 和 pET100/D – topo；用于聚合酶链反应（PCR）扩增的基因：pGFPuv 和质粒来源的 Kan 激酶基因。

（5）Vent 和 Taq DNA 聚合酶、限制性内切核酸酶、DNA 酶、T_4 激酶和 T_4 DNA 连接酶。

（6）寡核苷酸引物和 dNTP。

（7）谷胱甘肽琼脂糖球珠。

（8）L – 甲苯磺酰胺 – 2 – 苯乙基氯甲基酮处理过的胰蛋白酶。

3. 试剂

（1）细菌蛋白抽提物试剂（B – PER）和 B – PER Ⅱ。

（2）100 mmol 溶于水的异丙基 – β – D – 硫代半乳糖苷（IPTG）作为储备液。

（3）磷酸缓冲液（PBS）：10 × 储备液，每升含 80 g 氯化钠、2 g 氯化钾、11.5 g 磷酸氢二钠水合物（$Na_2HPO_4 \cdot 7H_2O$）和 2 g 磷酸二氢钾。

（4）1 mol/L 氯化镁。

（5）50 mmol/L Tris – HCl（pH 8.0）和 5 mmol/L 还原谷胱甘肽。

（6）Ni – NTA 纯化用的缓冲液：

①结合缓冲液，8 × 储备液：160 mmol/L Tris – HCl，pH 7.9，4 mol/L 氯化钠和 40 mmol/L咪唑。

②清洗缓冲液，8 × 储备液：160 mmol/L Tris – HCl，pH 7.9，4 mol/L 氯化钠和 480 mmol/L咪唑。

③洗脱缓冲液，8 × 储备液：160 mmol/L Tris – HCl，pH 7.9，4 mol/L 氯化钠和 2 mol/L咪唑。

（7）高效液相层析（HPLC）分析用的缓冲液：

①缓冲液 A：50 mmol/L NH_4OAc，pH 5.0。

②缓冲液 B：50 mmol/L NH_4OAc，pH 5.0 和 50% MeOH。

③缓冲液 C：0.1mol/L 磷酸二氢钠，pH 2.5。

④缓冲液 D：0.1mol/L 磷酸二氢钠，pH 2.5 和 50% MeOH。

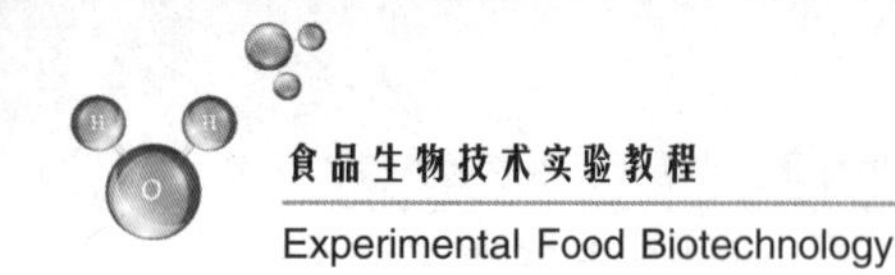

【实验步骤】

1. 大肠杆菌在含非天然氨基酸的培养基上生长

将大肠杆菌菌株 C600Δ*trp*E 及其衍生菌株接种于 M9B1TL 95% 4fW + Amp 培养基（即 M9 培养基补充了维生素 B_1、苏氨酸、亮氨酸、19∶1 的 4fW∶W 和 Amp）。培养过夜后，取适量培养液稀释 100 倍后，接种到 1.5 mL 的 LB 培养基中，以测试对生长的抑制。将这些 1.5 mL 的培养基按 350 μL/孔分到 3 个小孔中。于 37 ℃不停地摇动培养，直到所有培养物均进入稳定生长期。采用微板阅读器获得生长曲线。把生长曲线的指数部分拟合到以下方程，计算菌体固有的生长速率 r：

$$N_{(t)} = N_{(0)} \times e^{(-rt)}$$

式中，t 为时间；$N_{(t)}$ 为时刻 t 的菌体数（或光密度）；$N_{(0)}$ 是初始的菌体数。

2. 表达载体的构建

（1）pGSR。

由于 C600p 中的 pUC 18 质粒 pGEX - KG［一种谷胱甘肽 - S 转移酶（GST）表达载体］需要 Amp 以外的选择方法，因而，用 *Sma* Ⅰ和 *Eco*R Ⅰ降解质粒 pGEX - KG。P182Sfi - Kan 的 Kan 激酶基因由引物 Kan1.39（5’-CGCGGATCCGGCCACCATGGCCAAGCGAACCGGAAT）和 Kan2.39（5’-CGGAATTCTGAGGCCTGACAGGCCTTAGAAGAACTCGT）用 PCR 扩增。PCR 产物用 *Bsa*B Ⅰ和 *Eco*R Ⅰ降解并连接到 *Sma* Ⅰ和 *Eco*R Ⅰ降解后的质粒 pGEX - KG 上，然后转化到 $DH_{5\alpha}F'$ 中，再小量制备试剂盒分离质粒。得到的质粒 pGSR 是一种带有 Kan 抗性的 GST 表达载体。

（2）pET100GFPuv。

拓扑异构酶介导的定向克隆需要正向引物中的 5’-CACC 片段。用 Vent DNA 聚合酶（NEB）、引物 CF - PA（5’-CACCACGGCCACTGTGGCCATGAGTAAAGGAGAAGAACTT - 3）和 CF - PB（5’-GGCCATCGGGGCCCTATTTTATAGTTCATCCATGCC - 3’）通过 PCR 扩增质粒 pGFPuv 中的强荧光的 GFPuv 的基因。通过 7.5 μL PCR 产物与 1 μL 的 10×缓冲液、1 μL的 4 mmol dNTP 和 0.5 μL 的 Taq DNA 聚合酶在 72 ℃保育 20 分钟，把突出的腺苷残基加在扩增产物上。按说明书中指示，这个反应用来克隆 GFPuv 基因到 pET100/D - topo 中。拓扑异构酶反应是用来转化化学上具竞争性的 TOP10 细胞的。得到的质粒用小量制备试剂盒分离。PET100GFPuv 带有 Amp 抗性并且在 T_7 核酸（RNA）聚合酶启动子的控制下，需要在携带 λDE3 溶源体的寄主菌株中表达 GFPuv。

3. 高掺入 4fW 的 GST 的表达和纯化

（1）挑选在 M9B1TL 95% 4fW + Kan 培养基上过夜生长的、用 pGSR 转化的并在 Kan 板上选择的 C600p 的一单菌斑，接种于 M9B1TL 95% 4fW + Kan 液体培养基上，培养过夜。

（2）将上述培养物稀释100倍接种于100 mL同样基质的培养液中。把细菌培养到中指数期（mid-log）（在600 nm处，光密度0.5）。

（3）以5 400 转/分钟离心培养液20 分钟，并悬浮在补充了0.3 mmol/L IPTG的100 mL M9B1TL3 × 100% 4fW + Kan中，继续培养16小时。

（4）如步骤3将细胞离心，再用5 mL B－PER试剂裂解。

（5）离心去除不溶解部分之后，加入10 mmol/L 氯化镁（从1 mol储备液中取）和5U DNA酶，常温保育15分钟。

（6）加入500 μL 50%谷胱甘肽琼脂糖凝胶球珠浆到裂解液中，并在常温下用搅拌机混合2分钟。

（7）以最高速度简单离心球珠。加入5 mL PBS冲洗球珠。再重复冲洗2次。

（8）用1 mL冰冷的PBS做最后的冲洗，再移到微离心管中。

（9）在室温下用50 mmol/L Tris－HCl（pH 8.0），外加5 mmol/L还原的谷胱甘肽。在室温下旋转2分钟以从球珠上洗脱纯化的GST，分3次进行，每次0.5 mL。

（10）用Microcon浓缩器浓缩这3份样品。

（11）用SDS PAGE测定蛋白质纯度。纯度应高于95%。

4. 细胞全蛋白抽提物的纯化

（1）在25 mL恰当的基质中生长C600p到饱和。

（2）离心，去上清液，加200 μL B－PER裂解。

（3）取50 μL裂解液，用Centri－Sep尺寸排斥柱去掉未掺入的氨基酸，收集裂解液。

5. 纯化蛋白的分析

含非天然氨基酸的蛋白的水解和HPLC－ESI分析：

（1）冻干蛋白样品（如第4步中从Centri－Sep柱出来的半量洗脱液，或第3步中几毫克纯化蛋白）。

（2）将样品重悬在1 mL含10%巯基乙酸的5.4 mol/L氯化氢中以在水解中保护色氨酸。

（3）在真空和110 ℃下水解过夜。

（4）冻干水解产物，并重悬在50 μL的水中。

（5）用HPLC－ESI分析水解产物。当从HPLC柱上洗脱时，测定和跟踪天然的和非天然的氨基酸的比质量。用曲线下的面积确定洗脱质量的相对比率；用标准曲线确定氨基酸的实际摩尔数。

6. 蛋白酶降解以确定非天然氨基酸掺入水平

（1）冻干纯化的GST并重悬在0.1 mol/L碳酸氢铵中。

（2）在37 ℃用被固定化的L－甲苯磺酰胺－2－苯乙基氯甲基酮处理过的胰蛋白酶降解10小时。

（3）离心去除胰蛋白酶。

（4）冻干降解物，并加水重悬至 210 μmol/L。

（5）用 HPLC - ESI 分析降解产物，跟踪特定质量的洗脱剖面，并证明非天然氨基酸在特定片段中的掺入水平。

【注意事项】

（1）光学纯（对应体纯）的非天然氨基酸可能不易购得。如果是消旋混合物，应使用 20 μg/mL L - 对应体。如果非天然氨基酸是消旋混合物且要与天然氨基酸混合，应加倍非天然氨基酸的用量。例如，95% 4fW 需要 38 μg/mL DL - 4fW 加 1 μg/mL L - W。

（2）为了掺入非天然氨基酸，优先使用营养缺陷型菌株，但并非对所有应用都是必需的。

（3）此处讨论的是典型表达质粒的载体。一个是使用 T_7 RNA 聚合酶系统并在 *lac* 操纵子的控制之下；而另一个是为了表达目标基因则直接使用 *lac* 启动子。许多其他的表达载体已被用于把非天然氨基酸掺入感兴趣的蛋白质中，并用于完整的、部分的和定点的类似物掺入。

（4）为了达到最大的表达和目标基因的掺入，可能需要优化。优化的关键因素包括接种和转移到含引导物和非天然氨基酸的培养基之间的时机，以及表达延续时间的长短。例如，发现中指数期的培养液过夜表达的 GST 达到极大，而从中指数期再延续几小时的表达对 GFPuv 是最好的。更值得注意的是，用于荧光试剂的强蓝绿荧光蛋白的表达，包括荧光激活细胞分选，诱导过夜培养液再表达 6 小时的蛋白质最好。

（5）在这些例子中，通过检查产物中色氨酸的天然光吸收，简化用 HPLC 对水解蛋白的分析。如果对其他氨基酸类似物完成类似的实验，氨基酸在 HPLC 分析之前可能需要衍生化。

（6）在 0.45 μm 孔径的 Millipore 旋转柱中完成实验，对胰蛋白酶处理更有效。在反应的最后时间，胰蛋白酶可以用离心来纯化。固定化的胰蛋白酶吸附在膜中，洗脱液携带反应产物。

XⅢ. Whole Incorporation of Unnatural Amino Acids in *E. coli*

【Objectives】

(1) Understand the principle of incorporation of unnatural amino acids (tryptophan analogues) into *E. coli*.

(2) Be familiar with the experimental techniques of protein expression and purification.

(3) Understand the analysis method of the degree of the incorporation of unnatural amino acids into proteins.

【Principle】

The incorporation of tryptophan analogues into *E. coli* is straightforward. Because tryptophanyl tRNA synthetase has no editing domain, the difference between natural amino acids and analogues is only in structure. Tryptophanyl – tRNA synthetase from *Bacillus subtilis* can bind (load) fluorescent tryptophan analogues more effectively; the binding of 4-fluorotryptophan (4fW) is 6 times lower than that of tryptophan (W), while the binding of 5-fluorotryptophan (5fW) is 74 times lower than that of tryptophan. Similarly, tryptophan analogues have been known to enter human cells and support at least several generations of growth before the toxicity of the analogues takes effect. In order to effectively mix tryptophan analogues, mutant bacterial strains with the ability to prevent biosynthesis should be used.

【Apparatus, materials and reagents】

1. Apparatus

(1) Microcon concentrator, whose intercepted molecular weight is 10 000.

(2) Ni-Nitrotriacetic acid (Ni-NTA) resin (Novagen) and protein purification column.

(3) Centri-Sep size exclusion column.

(4) Agarose gel electrophoresis equipment for DNA and sodium dodecyl sulfate (SDS) gel

electrophoresis (PAGE) equipment.

(5) HPLC, HPLC-electrospray ionization (ESI) and mass spectrometry equipment.

(6) Microplate reader.

2. Materials

(1) *E. coli* strain C600Δ*trp*E (*thi*-1 *thr*-1 *leu*B6 *lac*Y1 *ton*A21 *sup*E44 *mcr*AΔ*trp*E) and its derivative strain C600p (C600 Δ*trp*E + pUC 18), C600pGSR (C600 Δ*trp*E + pG-SR), C600F (C600 Δ*trp*E F' KanR) and C600F (DE3) (C600FλDE3 lysate). $DH_{5\alpha}$ F' and TOP10 are used for transformation.

(2) Luria-Bertani medium (10 g tryptone, 5 g yeast extract, 10 g sodium chloride and 1.5% agarose for bacteria for culture plate per liter) and basic medium M9 (5 × stock solution: 30 g disodium hydrogen phosphate, 15 g potassium dihydrogen phosphate, 5 g ammonium chloride, 2.5 g sodium chloride and 1.5% agarose for bacteria for culture plate per liter) are supplemented with 20 μg/mL amino acids and 0.000 5% thiamine. The richment medium and basic medium are supplemented with antibiotics as indicated: 50 μg/mL ampicillin (Amp) or kanamycin (Kan).

(3) Tryptophan analogues: 4-fluorotryptophan (4fW).

(4) Plasmids for high protein expression: pGEX-KG and pET100/D-topo. Gene for polymerase chain reaction (PCR) amplification: pGFPuv and plasmid derived Kan kinase gene.

(5) Vent and Taq DNA polymerase, restriction endonuclease, DNase, T_4 kinase and T_4 DNA ligase.

(6) Oligonucleotide primers and dNTP.

(7) Glutathione-agarose beads.

(8) Trypsin treated with L-toluenesulfonylamine-2-phenylethyl chloromethyl ketone.

3. Reagents

(1) B-PER and B-PER Ⅱ for extracting bacterial proteins.

(2) Dissolve 100 mmol isopropyl-β-D-galactoside (IPTG) in water as the stock solution.

(3) Phosphate buffer (PBS): 10 × stock solution: 80 g sodium chloride, 2 g potassium chloride, 11.5 g disodium hydrogen phosphate hydrate ($Na_2HPO_4 \cdot 7H_2O$) and 2 g potassium dihydrogen phosphate per liter.

(4) 1 mol/L magnesium chloride.

(5) 50 mmol /L Tris-HCl (pH 8.0) and 5 mmol/L reduced glutathione.

(6) Buffer for Ni-NTA purification:

① Binding buffer, 8 × stock solution: 160 mmol/L Tris-hydrogen chloride, pH 7.9, 4 mol/L sodium chloride and 40 mmol/L imidazole.

② Washing buffer, 8 × stock solution: 160 mmol/L Tris-hydrogen chloride, pH 7.9, 4 mol/L sodium chloride and 480 mmol/L imidazole.

③ Elution buffer, 8 × stock solution: 160 mmol/L Tris-hydrogen chloride, pH 7.9, 4 mol/L sodium chloride and 2 mol/L imidazole.

(7) Buffer for HPLC analysis.

①Buffer A: 50 mmol/L NH_4OAc, pH 5.0.

②Buffer B: 50 mmol/L NH_4OAc, pH 5.0 and 50% MeOH.

③Buffer C: 0.1 mol/L sodium dihydrogen phosphate, pH 2.5.

④Buffer D: 0.1 mol/L sodium dihydrogen phosphate, pH 2.5 and 50% MeOH.

【Procedures】

1. Growth of *E. coli* on medium with unnatural amino acids

The *E. coli* Strain C600 Δ*trp*E and its derivatives are inoculated in liquid M9B1TL 95% 4fW + Amp medium (that is, M9 medium supplemented with thiamine, threonine, leucine, 19 : 1 of 4fW : W and Amp). After overnight culture, dilute 100 times with appropriate medium, and inoculate it into each 1.5 mL LB medium to test the inhibition of growth. The 1.5 mL medium is divided into three cavities, 350 μL/cavity. Cultivate it with shaking at 37 ℃ until all cultures enter the stationary period. The growth curve is obtained by using the microplate reader. The exponential part of the growth curve is fitted to the following equation to calculate the inherent growth rate r:

$$N_{(t)} = N_{(0)} \times e^{(-rt)}$$

In the formula, t is time; $N_{(t)}$ is the cell number (or optical density) at time t; $N_{(0)}$ is the initial cell number.

2. Construction of expression vector

(1) pGSR.

Because the pUC 18 plasmid pGEX-KG [a glutathione s transferase (GST) expression vector] in C600p needs a selection method other than Amp, *Sma* Ⅰ and *Eco*R Ⅰ are used to degrade the plasmid. The Kan kinase gene of P182Sfi-Kan is amplified by PCR with primers Kan1.39 (5'-CGCGGATCCGGCCACCATGGCCAAGCGAACCGGAAT) and Kan2.39 (5'-CGGAATTCTGAGGCCTGACAGGCCTTAGAAGAACTCGT). The PCR products are degraded by *Bsa*B Ⅰ and *Eco*R Ⅰ and ligated to the plasmid pGEX-KG degraded by *Sma* Ⅰ and *Eco*R Ⅰ, then transformed into $DH_{5\alpha}$ F'. The plasmids are isolated by kits prepared in small quantities. The obtained plasmid pGSR is a GST expression vector with Kan resistance.

(2) pET100GFPuv.

Topoisomerase mediating directed cloning needs 5' -CACC fragment in forward primer. The gene of GFPuv with high fluorescence is amplified by PCR from the plasmid pGFPuv with vent DNA polymerase (NEB), primers CF-PA (5' -CACCACGGCCACTGTGGCCATGAGTAAAG-GAGAAGAACTT-3) and CF-PB (5' -GGCCATCGGGGCCCTATTTATAGTTCATCCATGCC-3'). 7.5 μL PCR product is incubated with 1 μL 10 × buffer, 1 μL 4 mmol dNTP and 0.5 μL Taq DNA polymerase at 72 ℃ for 20 mins. The prominent adenosine residues are added to the PCR product. According to the manufacturer's instructions, this reaction is used to clone the GF-Puv gene into pET100/D-topo. The reaction of topoisomerase is used to transform chemically competitive TOP10 cells. The plasmids are isolated by kits prepared in small quantities. PET100GFPuv is Amp resistant and GFPuv needs to be expressed in the host strain carrying λDE3 lysate under the control of T_7 RNA polymerase promoter.

3. Expression and purification of GST with high-content 4fW incorporation

(1) A single plaque of C600p, which grows overnight on M9B1TL 95% 4fW + Kan medium, transformed with pGSR and selected on Kan plate, is inoculated on M9B1TL 95% 4fW + Kan liquid medium and cultured overnight.

(2) The above culture is diluted 100 times and inoculated in 100 mL of the same medium. The bacteria are cultured to mid-log (optical density 0.5 at 600 nm).

(3) The medium is centrifuged at 5 400 rpm for 20 mins. The cells are suspended in 100mL M9B1TL3 × 100% 4fW + Kan medium supplemented with 0.3 mmol/L IPTG to culture for 16 h.

(4) The culture is then centrifuged as in step (3), and the cells are collected and lysed with 5 mL of B-PER.

(5) Centrifuge to collect the supernatant. Add 10 mmol/L magnesium chloride (taken from 1 mol stock solution) and 5U DNA polymerase, incubate at room temperature for 15 mins.

(6) Add 500 μL of 50% glutathione-agarose gel beads to the lytic solution and mix with 2 mins stirring at room temperature.

(7) Centrifuge at the highest speed and collect the beads. Add 5 mL PBS to rinse the beads. Rinse again for 2 times.

(8) The final rinse is done with 1 mL cold PBS and then the beads are transferred to a micro centrifuge tube.

(9) At room temperature, 50 mmol/L Tris-hydrogen chloride (pH 8.0) and 5 mmol/L reduced glutathione are added. The purified GST is eluted three times from the beads by rotating at room temperature for 2 mins with 0.5 mL each time.

(10) The three samples are concentrated in Microcon concentrator.

(11) The purity of protein is determined by SDS PAGE, and the purity should be higher

than 95%.

4. Purification of cell protein extract

(1) C600p is inoculated in 25 mL of medium to grow to saturation.

(2) Centrifuge the culture and collect the cells. Lyse the cells with 200 μL B-PER;

(3) Apply the Centri-Sep size exclusion column to remove the amino acids with 50 μL of lytic solution. And collect the lytic solution.

5. Analysis of purified protein

Hydrolysis and HPLC-ESI analysis of proteins containing unnatural amino acids:

(1) Lyophilized protein samples (e. g. , half of the eluent from the Centri-Sep column in step 4 or a few milligrams of purified protein in step 3) .

(2) The sample should be resuspended in 1 mL of 5. 4 mol/L hydrogen chloride containing 10% mercaptoacetic acid to protect tryptophan during hydrolysis.

(3) Hydrolysis should be carried out overnight under vacuum and at 110 ℃.

(4) The hydrolysate is lyophilized and resuspended in 50 μL water.

(5) The hydrolysates are analyzed by HPLC-ESI. When eluted from the HPLC column, the specific mass of natural and unnatural amino acids is determined and tracked. The area under the curve is used to determine the relative ratio of elution mass and the standard curve is used to determine the actual mole number of amino acids.

6. Protease degradation to determine the incorporation level of unnatural amino acids

(1) The purified GST is lyophilized and resuspended in 0. 1 mol/L ammonium bicarbonate.

(2) Trypsin treated with immobilized L-toluenesulfonamide-2-phenylethyl chloromethyl ketone is used to degrade GST at 37 ℃ for 10 h.

(3) Remove trypsin by centrifugation.

(4) The degradation product is lyophilized and resuspended to 210 μmol/L with water.

(5) Analyze the degradation product by HPLC-ESI, track the elution profile of specific mass, and prove the incorporation level of unnatural amino acids in specific fragments.

【Notes】

(1) Unnatural amino acids of optical purity (homologous purity) may not be readily available. In the case of racemic mixture, 20 μg/mL L-counterpart should be used. If the unnatural amino acid is a racemic mixture and is to be mixed with natural amino acid, the amount of unnatural amino acid should be doubled. For example, 95% 4fW requires 38 μg/mL DL-4fW plus 1 μg/mL L-W.

(2) To mix unnatural amino acids, priority should be given to the use of nutrient deficient

strains, but it is not necessary for all applications.

(3) The vector of typical expression plasmid is discussed here. One is to use T_7 RNA polymerase system under the control of *lac* operon; the other is to use *lac* promoter directly to express target gene. Many other expression vectors have been used for the incorporation of unnatural amino acids into target proteins and for the incorporation of complete, partial and site-specific analogues.

(4) To achieve the maximum expression and the incorporation of target genes, optimization may be needed. The key factors of optimization include the timing of inoculation and transfer to the medium containing guide and unnatural amino acids, and the duration of expression. For example, it is found that the GST expressed overnight in medium exponential phase reaches the maximum, and the expression lasting for several hours from medium exponential phase is the best for GFPuv. What's more, strong blue-green fluorescent protein used for fluorescent reagents, including fluorescence activated cell sorting, is the best protein for inducing overnight culture medium to express again for 6 hours.

(5) In these cases, the analysis of hydrolyzed protein by HPLC is simplified by examining the natural absorption of photo tryptophan in the product. If similar experiments are performed on other amino acid analogues, amino acids may need to be derivatized before HPLC analysis.

(6) It is more effective for trypsin treatment to complete the experiment in Millipore rotation column with pore size of 0.45 μm. At the end of the reaction, trypsin can be purified by centrifugation. The immobilized trypsin is adsorbed on the membrane, and the eluent carried the reaction product.

十四、人造锌指蛋白的设计和合成

【实验目的】

（1）了解锌指蛋白的设计原理和设计技巧。

（2）熟悉人工锌指蛋白的构建步骤。

【实验原理】

$(Cys)_2$ $(His)_2$型锌指模体（真核细胞中最常见的 DNA 结合模体）为新的 DNA 结合蛋白设计提供通用且易用的框架。一些 $(Cys)_2$ $(His)_2$型锌指蛋白 - DNA 复合物反映了这一模体与 DNA 结合具有下列特征：

（1）锌指结构用一特异的连接子反复连接。

（2）每一个锌指结构结合到 DNA 的一个三碱基对位置上，以其螺旋对着主沟槽。

（3）a 螺旋的每一锌指单元的 4 个关键位置上的氨基酸残基与 DNA 碱基在特定位置上 1：1 地接触。

（4）（锌指）多肽的整体排列是与 DNA 主要作用链反向平行的。

基于这些特征，科学家们成功地制造了某些锌指蛋白，如长序列或 AT 序列结合锌指。这样设计的 DNA 结合蛋白有望是拥有具高亲和力和特异性的唯一结合序列。

【实验仪器、材料与试剂】

1. 仪器

DNA 测序仪、超声裂解设备、十二烷基硫酸钠聚丙烯酰胺凝胶电泳（SDS PAGE）设备、色谱设备、高 S 色谱柱、UNO - S1 色谱柱、9 - 芴甲氧羰基固相合成设备、高效液相色谱、μBonedespheres C4 - 300（19 mm × 150 mm）柱、飞行时间质谱、圆二色（CD）谱设备、紫外—可见（UV - VIS）光谱仪、核磁共振仪（NMR）。

2. 材料

pET3b 表达系统、Superdex 75、互补 DNA 编码 Sp1 锌指、正文中描述的寡核苷酸引物、限制酶、T_7 DNA 聚合酶、T_4 DNA 连接酶。

Luria - Bertani（LB）培养基：将 10 g 胰蛋白胨、5 g 酵母抽提物和 10 g 氯化钠放入锥形烧瓶中，加水至终体积1L。

3. 试剂

大肠杆菌菌株$DH_{5\alpha}$和BL21（DE3）pLysS、氨苄西林、异丙基－β－D硫代－吡喃半乳糖苷（IPTG）、130 mmol/L 氯化钠、2.7 mmol/L 氯化钾、50 mmol/L Tris－HCl（pH 8.0）、50 mmol/L 氯化钠、1 mmol/L 二硫苏糖醇和三氟乙酸：乙醇（95：5）溶液。

磷酸缓冲生理盐水（PBS）：10 mmol/L 磷酸缓冲液（pH 7.6）。

【实验步骤】

1. 人工锌指蛋白的表达和纯化

（1）pEV－3b表达载体。

①从pET3b中切除*Eco*R Ⅰ/*Hin*d Ⅲ降解片段。

②将退火的寡核苷酸5’－AATTGTCATGTTTGAC－3’和5’－AGCTGTCAAACATGAC－3’插入*Eco*R Ⅰ/*Hin*d Ⅲ降解的pET3b，产生的质粒记为pET3b’。

③制备包括限制性酶*Afl*I Ⅰ、*Bam*H Ⅰ、*Eco*R Ⅰ、*Hin*d Ⅲ和*Sma* Ⅰ酶切位点的双链寡核苷酸。这两条链的序列为5’－TATGGATCCCGGGAATTCAAGCTTAAGC－3’和5’－TCAGCTTAAGCTTGAATTCCCGGGATCCA－3’。

④用*Nde* Ⅰ和*Bpu*1102 Ⅰ降解，并将这一片段插入pET3b’，得到质粒pEV－3b。得到的质粒pEV－3b用来表达设计的锌指蛋白。把锌指基因片段插入*Bam*H Ⅰ/*Eco*R Ⅰ降解的pEV－3b，只用一步就构建了人工锌指的表达载体。

（2）设计的人工锌指蛋白的表达。

①把带有设计的锌指蛋白基因的pEV－3b质粒转化到大肠杆菌菌株BL21（DE3）pLysS中。

②在37 ℃于LB培养基中培养细胞。

③当在600 nm的光密度为0.6时加入1 mmol/L IPTG。

④培养液在20 ℃保育8～12小时。为了表达溶解状态的蛋白质，这个温度很重要。

（3）设计的人工锌指蛋白的纯化。

纯化步骤在4 ℃下完成。

①收获大肠杆菌细胞，重悬，并在PBS中裂解。

②离心后，用阳离子交换色谱（高S和UNO S－1）和用Tris－HCl缓冲液凝胶过滤（Superdex 75），纯化溶有锌指蛋白的上清液。

③用SDS－PAGE确认蛋白纯度。

为进行金属取代实验，不溶形式的锌指蛋白也在离心后从大肠杆菌细胞小团中纯化出来：

①在含8 mmol/L尿素和10 mmol/L螯合剂（EDTA或1，10－邻二氮杂菲）的PBS中

裂解细胞小团。

②根据（3）中第②步的描述，以同样步骤纯化。

③在 65 ℃加热纯化蛋白 30 分钟，并在含 125 μmol/L 氯化锌、硝酸镍、氯化镉、硝酸钴或硫酸铜的 10 mmol/L Tris - HCl 缓冲液中逐步降温，使之重新折叠。

（4）CD 测量。

用 Jasco J - 720 分光偏振计记录锌指蛋白的 CD 谱。20 ℃，使用 pH 8.0 的含有 50 mmol/L氯化钠的 Tris - HCl 缓冲液。容器须有盖，1 mm 光程，氮气环境。所有的谱平均扫描 8 ~16 次。使用 Jasco 软件校准谱的基线并降低噪声。

（5）UV - VIS 吸收谱。

用 Beckman Couter DU7400 二极管阵列光谱仪测 UV - VIS 吸收谱，20 ℃，使用pH 7.5 的含 50 mmol/L 氯化钠的 10 mmol/L Tris - HCl 缓冲液。容器须有盖，1 cm 光程。Co（Ⅱ）取代的锌指复合物用氯化钴滴定得到。多肽在任意条件下用 Co（Ⅱ）饱和。所有的谱都用 $\varepsilon = A/(l \cdot c)$，式中 ε 是消光系数［（mol/L）/cm］，$l$ 是容器的光程（cm），c 是肽浓度（mol/L）。

（6）NMR 实验。

在 1.5 mol Zn（Ⅱ）离子存在的条件下，单指结构域和 Zn（Ⅱ）的复合物在 90% H_2O/10% D_2O 和 D_2O（25 mol/L Tris - d_{11}，pH 5.7）中，以 5 mol/L 浓度制备。所有的 NMR 谱都在 JOEL Lambda - 600 谱仪上记录。

①核 Overhauser 增强谱（NOESY）数据的采集：先用选择性水预饱和，随后温度定在 30 ℃，各使用混合时间分别为 100 毫秒、200 毫秒和 300 毫秒的标准 NOESY 脉冲序列。

②采集总相关谱，用 80 毫秒 MLEV - 17 自旋锁闭时长，在 30 ℃用梯度修整脉冲以压制水信号。

谱的典型采集条件是每个 t_1 值扫描 24 次，共 1 024 个 t_1 值，并且在直接维度收集2 048 个复数点。在两个维度的自由感应衰减被乘以相移正弦钟形限形函数、零填充，并经傅里叶变换为 2 048 ×2 048 矩阵。序列共振指认用标准的总相关谱和 NOESY 过程确定。

2.6 和 9 锌指蛋白

（1）蛋白质设计策略。

全新的 6 和 9 锌指蛋白（Sp1ZF6 和 Sp1ZF9）是从转录因子 Sp1 的 3 锌指模体创造出来的。通过用 Krüppel 型连接子连接 2 或 3 个 Sp1 锌指结构域，构建出了这些蛋白质。

（2）Sp1ZF6 和 Sp1ZF9 基因的构建。

编码 Sp1 的 3 锌指区域的基因［pUC - Sp1（530 ~623）］按先前描述过的步骤构建。

①合成编码 Krüppel 型连接子（TGEKP）的段寡核苷酸（84 bp）为 *Bam*H Ⅰ］/*Sty* Ⅰ片段，并将其插入 pUC - Sp1（530 ~623）。

②切出 *Eco* 47 Ⅲ片段（264 bp），并将其插入简单降解过的 pUC - Sp1（530 ~623）

中。改变后的质粒重新命名为 pUC－Sp1ZF6。

③用带有 *Age* Ⅰ酶切位的引物对，即用 5’－ACCGGTGAAAAACCGCATATTTGCCA-CATAC－3’为编码链和 5’－CGGTTTTTCACCGTGTGGGTCTTGATATG－3’为非编码链，*Age* Ⅰ酶切后连接到编码 Sp1ZF9 的中间 Sp1 基因的 5’端和 3’端。

④把得到的用 *Age* Ⅰ修饰过的片段连接到 pUC－Sp1ZF6 的 *Age* Ⅰ位置。连接 2 或 3 个 Sp1 片段间的 *Age* Ⅰ酶切位点编码氨基酸 TG，它是连接肽 TGEKP 的一部分。

⑤用 DNA 测序确认所有序列。

⑥作为 *Bam*H Ⅰ/*Sty* Ⅰ片段切出 Sp1ZF6 和 Sp1ZF9 的 DNA 片段，并将其插入简单降解过的质粒 pEV－3b 中［见 1 中（1）］。

【注意事项】

（1）Krüppel 型连接子（Thr－Gly－Glu－Lys－Pro，TGEKP）在许多锌指蛋白中是保守的，因而被选来连接 Sp1 的锌指结构域。这些人工多锌指蛋白显示出扩展了的序列特异性，并且它们对序列的偏好取决于模体数量和 Sp13 锌指 DNA 结合结构域的特征。

（2）显然，足迹分析证明，现有的人工锌指蛋白中 Sp1ZF6 和 Sp1ZF9 使用它们所有的锌指结构域分别与 DNA 序列中的最少 18 和 27 个连续的 GC 富集碱基对结合。

（3）新设计的 6 锌指蛋白 Sp1ZF6（Gly）$_7$和 Sp1ZF6（Gly）$_{10}$能够在两个末端的结合部位交界处引起 DNA 弯折，并且，两个 3 锌指模体间连接子的长度对整个 DNA 弯折的方向起关键作用。相检测强烈地支持所引起的 DNA 弯折朝向 DNA 的主沟槽，并且，Sp1ZF6（Gly）$_7$引起 DNA 弯折向最强烈的方向改变。

（4）通过用不同的带电连接子连接 Sp1 的两个 DNA 结合结构域，我们也可以制造出 6 锌指蛋白 Sp1ZF6（Gly·Arg）$_4$和 Sp1ZF6（Gly·Glu）$_4$。

XIV. Design and Synthesis of Artificial Zinc Finger Protein

【Objectives】

(1) Understand the design principle and skills of zinc finger protein.

(2) Be familiar with the construction steps of artificial zinc finger protein.

【Principle】

The $(Cys)_2$ $(His)_2$zinc finger motifs (the most common DNA binding motif in eukaryotic cells) provide a universal and attractive framework for the design of new DNA binding proteins. Some $(Cys)_2$ $(His)_2$ type zinc finger protein-DNA complexes have revealed the following characteristics of the binding of DNA to the motif:

(1) The zinc finger structure is repeatedly connected by a specific linker.

(2) Each zinc finger structure binds to a three base pair position of DNA with its helix facing the main groove.

(3) The amino acid residues at four key positions of each zinc finger unit of a-helix have 1 : 1 contact with DNA bases at specific positions.

(4) The overall arrangement of (zinc finger) polypeptides is in reverse parallel with the main acting strand of DNA.

According to these characteristics, scientists have successfully produced some zinc finger proteins, such as long sequence or AT sequence binding zinc finger. The designed DNA binding protein can be expected to have a unique binding sequence with high affinity and specificity.

【Apparatus, materials and reagents】

1. Apparatus

DNA sequencer, ultrasonic equipment, sodium dodecyl sulfate-polyacrylamide gel electrophoresis (SDS PAGE) equipment, chromatographic equipment, high S column, UNO-S1 column, equipment for solid phase synthesis of 9-fluorenylmethoxycarbonyl, high performance liquid chromatography, μBonedespheres C4-300 (19 mm × 150 mm) column, time-of-flight mass spectrometry, circular dichroism (CD) spectrum, UV VIS spectrometer and nuclear magnetic resonance instrument (NMR).

2. Materials

pET3b expression system, Superdex 75, complementary DNA encodes Sp1 zinc finger, oligonucleotide primers described in the text, restriction enzyme, T_7 DNA polymerase, T_4 DNA ligase.

Luria-Bertani (LB) medium: take 10 g tryptone, 5 g yeast extract and 10 g sodium chloride in an Erlenmeyer flask. Add water to the final volume of 1 L.

3. Reagents

E. coli strains $DH_{5\alpha}$, BL21 (DE3) pLysS, Ampicillin, isopropyl-β-D thiogalactopyranoside (IPTG), 130 mmol/L sodium chloride, 2.7 mmol/L potassium chloride, 50 mmol/L Tris-hydrogen chloride (pH 8.0), 50 mmol/L sodium chloride, 1 mmol/L dithiothreitol and trifluoroacetic acid : ethanol (95 : 5) solution.

Phosphate buffered normal saline (PBS): 10 mmol/L phosphate buffer (pH 7.6).

【Procedures】

1. Expression and purification of artificial zinc finger protein

(1) pEV-3b expression vector.

①The *Eco*R Ⅰ/*Hin*d Ⅲ degradation fragments is excised from pET3b.

②Annealed oligonucleotides 5'-AATTGTCATGTTTGAC-3' and 5'-AGCTGTCAAACATGAC-3' are inserted into *Eco*R Ⅰ/*Hin*d Ⅲ degraded pET3b. The resulting plasmid is named pET 3b'.

③Double stranded oligonucleotides including restriction enzyme *Afl*I Ⅰ, *Bam*H Ⅰ, *Eco*R Ⅰ, *Hin*d Ⅲ and *Sma* Ⅰ are prepared. The sequences of the two chains are 5'-TATGGATCCCGGGAATTCAAGCTTAAGC-3' and 5'-TCAGCTTAAGCTTGAATTCCCGGGATCCA-3'.

④The above fragment plasmid is degraded by *Nde* Ⅰ and *Bpu*1102 Ⅰ and then inserted into pEV3b' to obtain pEV-3b which is designed to express zinc finger protein. Zinc finger gene fragment is inserted into *Bam*H Ⅰ/*Eco*R Ⅰ degraded pEV-3b, and the expression vector of artificial zinc finger is constructed in one step.

(2) Expression of artificial zinc finger protein.

①The plasmid of pEV-3b with the designed zinc finger protein gene is transformed into BL21 (DE3) pLysS.

②The cells are cultured in LB medium at 37 ℃.

③When the optical density of 600 nm is 0.6, 1 mmol/L IPTG is added.

④The medium is kept at 20 ℃ for 8 – 12 hours. This temperature is important to express the protein in the dissolved state.

(3) Purification of artificial zinc finger protein.

The purification procedure should be completed at 4 ℃.

①The *E. coli* cells are harvested, suspended and lysed in PBS.

②After centrifugation, the supernatant of zinc finger protein is purified by cation exchange chromatography (high S and UNO S-1), and gel filtration (Superdex 75) with Tris-hydrogen chloride buffer.

③SDS-PAGE is used to confirm the purity of the protein.

To carry out metal substitution experiments, the insoluble zinc finger protein is purified from the small cell mass of *E. coli* after centrifugation:

①The small cell clusters are lysed in PBS containing 8 mmol/L urea and 10 mmol/L chelating agent (EDTA or 1, 10-phenanthroline).

②Purification in the same step as described in step ② of subsection (3).

③The protein is purified for 30 mins at 65 ℃, and then cooled gradually in 10 mmol/L Tris-hydrogen chloride buffer containing 125 μmol/L zinc chloride, nickel nitrate, cadmium chloride, cobalt nitrate or copper sulfate, so as to fold it again.

(4) CD measurement.

CD spectrum of zinc finger protein is recorded by Jasco J-720 spectropolarimeter. At 20 ℃, Tris-hydrogen chloride buffer containing 50 mmol/L sodium chloride at pH 8. 0 is used. The vessel must have a cover and 1 mm light path with nitrogen environment. All the spectra are scanned 8 – 16 times on average. Use Jasco software to calibrate baseline and reduce noise.

(5) UV-VIS absorption spectrum.

The absorption spectrum of UV-VIS is determined by Beckman Couter DU7400 diode array spectrometer, 20 ℃, and 10 mmol/L Tris-hydrogen chloride buffer containing 50 mmol/L sodium chloride (pH 7. 5) is used. The container must have a cover and 1 cm light path. The zinc finger complex replaced by Co (Ⅱ) is titrated by cobalt chloride. The polypeptide is saturated with Co (Ⅱ) under any condition. All spectrum are $\varepsilon = A/(l \cdot c)$, where ε is extinction coefficient [(mol/L) /cm], l is the light path of the container (cm), and c is the peptide concentration (mol/L).

(6) NMR experiment.

In the presence of 1. 5 mol Zn (Ⅱ) ions, single finger domain and the complex of Zn(Ⅱ) are prepared in 90% H_2O/10% D_2O and D_2O (25 mol/L Tris-d_{11}, pH 5. 7) to 5 mol/L. All NMR spectra are recorded on JOEL Lambda-600 spectrometer.

①The data of nuclear Overhauser enhanced spectroscopy (NOESY) are collected by using the standard NOESY pulse sequences by pre-saturation with selective water at 30 ℃, mixing time of 100 ms, 200 ms and 300 ms, respectively.

②Collect total correlation spectrums, and use the time of 80 ms MLEV-17 spin locking, and the gradient trimming pulse to suppress the water signal at 30 ℃.

The typical collecting conditions of spectrum are that each t_1 value is scanned 24 times, with a total of 1 024 t_1 values, and 2 048 complex points are collected in the direct dimension. The free induction attenuation in two dimensions is multiplied by the phase shift sinusoidal bell shaped limiting function and zero filling, and is transformed into 2 048 × 2 048 matrix by Fourier transform. The sequence resonance is determined by the standard total correlation spectrum and NOESY process.

2. 6 and 9 zinc finger proteins

(1) Protein design strategy.

The new 6 and 9 zinc finger proteins (Sp1ZF6 and Sp1ZF9) were created from the 3 zinc finger models of transcription factors Sp1. These proteins are constructed by using a linker of Krüppel type to connect 2 or 3 Sp1 zinc finger domains.

(2) Construction of Sp1ZF6 and Sp1ZF9 genes.

The gene encoding the 3 zinc finger region of Sp1 [pUC – Sp1 (530 – 623)] is constructed as described previously.

①The synthesized 84 bp fragment encoding the Krüppel type linker (TGEKP) is a *Bam*H Ⅰ/*Sty* Ⅰ fragment. Inserted it into pUC – Sp1 (530 – 623).

②The *Eco* 47 Ⅲ fragment (264 bp) is cut and inserted into pUC – Sp1 (530 – 623). The modified plasmid is renamed as pUC-Sp1ZF6.

③Use 5'-ACCGGTGAAAAACCGCATATTTGCCACATAC-3' as a coding chain and 5'-CGGTTTTTCACCGTGTGGGTCTTGATATG-3' as a noncoding chain. Connect the two *Age* Ⅰ sites to the 5' and 3' ends of the intermediate Sp1 gene encoding Sp1ZF9, respectively.

④The obtained fragment modified with *Age* Ⅰ is connected to the *Age* Ⅰ site of pUC-Sp1ZF6. The amino acid TG is encoded by an *Age* Ⅰ restriction site between two or three Sp1 fragments, which is a part of the linker peptide TGEKP.

⑤All sequences are confirmed by DNA sequencing.

⑥As *Bam*H Ⅰ/*Sty* Ⅰ fragments, the DNA fragments of Sp1ZF6 and Sp1ZF9 are cut and inserted into the plasmid pEV-3b, see (1) in 1.

【Notes】

(1) Krüppel type linker (Thr-Gly-Glu-Lys-Pro, TGEKP) is conserved in many zinc finger

proteins, so it is selected to connect the zinc finger domain of Sp1. These artificial zinc finger proteins show extended sequence specificity, and their preference for sequence depends on the number of motifs and the characteristics of Sp13 zinc finger DNA binding domain.

(2) Obviously, footprint analysis showed that Sp1ZF6 and Sp1ZF9 used all their zinc finger domains to bind to at least 18 and 27 consecutive GC enriched base pairs in DNA sequence, respectively.

(3) The newly designed 6-zinc finger proteins Sp1ZF6 $(Gly)_7$ and Sp1ZF6 $(Gly)_{10}$ can cause DNA bending at the junction of the two terminal binding sites, and the length of the linker between the two 3-zinc finger motifs plays a key role in the direction of the whole DNA bending. Phase detection strongly supports that the resulting DNA bending is toward the main groove of DNA, and Sp1ZF6 $(Gly)_7$ causes the strongest change in the direction of DNA bending.

(4) 6-zinc finger proteins Sp1ZF6 $(Gly \cdot Arg)_4$ and Sp1ZF6 $(Gly \cdot Glu)_4$ are created by connecting two DNA binding domains of Sp1 with different charged linkers.

十五、位点特异性核酸内切酶的蛋白质工程

【实验目的】

（1）学习改变位点特异性核酸内切酶特异性设计项目中所涉及的各种步骤。

（2）学会分析蛋白质结构变化的方法。

（3）学会鉴别负责感知甲基化状态的候选氨基酸残基。

【实验原理】

转换错配修复核酸酶 MutH 被 MetS 和 MutL 刺激的时候，会在半甲基化位点 d（GATC）切开一缺口，转化为切割全甲基化以及半甲基化和未甲基化 DNA 的变体。

【实验软件、仪器、材料与试剂】

1. 软件

ClustalW（http：//www. ebi. ac. uk/clustalw/）、ClustalX（http：//bips. u - strasbg. fr/fr/Documentation/ClustalX/）、GeneDoc（http：//www. psc. edu/biomed/genedoc/）、RasMol（http：//www. openrasmol. org/）、Swiss PDB Viewer（http：//www. expasy. org/spdbv/）。

2. 仪器

超声破碎器、冷冻离心机、空柱、GeneScan－500 TAMRA DNA 分子质量标准、ABI PRISM 310 基因分析仪。

3. 材料

pMQ402、PCR 引物、pfu 聚合酶、脱氧核糖核苷三磷酸、阿拉伯糖、苯甲基酰氟（PMSF）、Ni－NTA 琼脂糖、大肠杆菌菌株 XL－1 Blue MRF、大肠杆菌 MutL 蛋白、荧光团标记的寡核苷酸或荧光标记的 PCR 产物、模板 DNA、磷酸化引物光团标记引物、dam 甲基化酶、λ－外切酶、外切酶、P 旋柱和 pfu DNA 聚合酶、灌有加了 8 mol/L 尿素的 POP－4 聚合物的 47 cm 毛细管（内径 50 μm）。

4. 试剂

（1）LB 培养基（每升）：10 g 菌用胰蛋白胨、5 g 菌用酵母抽提物和 5 g 氯化钠。用氢氧化钠调节 pH 至 7.2～7.5，灭菌。

（2）结合缓冲液：20 mmol/L Tris - HCl、5 mmol/L 咪唑、1 mol/L 氯化钠和1 mmol/L PMSF（pH 7.9）。

（3）清洗缓冲液：20 mmol/L Tris - HCl、60 mmol/L 咪唑、1 mol/L 氯化钠和 1 mmol/L PMSF，pH 7.9。

（4）洗脱缓冲液：20 mmol/L Tris - HCl、200 mmol/L 咪唑、1 mol/L 氯化钠和 1 mmol/L PMSF（pH 7.9）。

（5）透析缓冲液：10 mmol/L HEPES 氢氧化钾、100 mmol/L 氯化钾、1 mmol/L EDTA 和 1 mmol/L 二硫苏糖醇（pH 7.9）。

（6）透析缓冲液 G：10 mmol/L HEPES 氢氧化钾、100 mmol/L 氯化钾、1 mmol/L EDTA、1 mmol/L 二硫苏糖醇和 50% 甘油，pH 7.9。

（7）测试缓冲液：10 mmol/L Tris - HCl（pH 7.5）、10 mmol/L 氯化镁、0.75 mmol/L 三磷酸腺苷和 0.1 mg/mL 牛血清白蛋白。

（8）加有 1 mmol/L EDTA 的 1 × 基因分析缓冲液。

（9）模板抑制剂。

【实验步骤】

（1）从生物技术信息国家中心基因组基本局部比对搜索工具（basic local alignment search tool，BLAST）页面（http：//www. ncbi. nlm. nikh go/BLAST）得到 MutH 蛋白和相关限制性内切核酸酶的序列。更多的序列从 PSI BLAST 服务器得到。使用 PAM250 矩阵，用 ClustalX 程序比对序列，比对好的序列用 GeneDoc 程序分析。程序的使用方法可在程序内使用“Help”功能得到。结构的坐标从 PDB 数据库（http：//www. resb. org/pdb/）得到。程序 RasMol 和 Swiss PDB Viewer 用于结构观察。

（2）将比对结果输入 GeneDoc 程序，用“Group”功能将序列分组（如一个组包含 MutH，而另一个组包含限制性内切核酸酶），鉴别组特异的残基。用 GeneDoc 内的“RasMol Script Dialog”功能生成 RasMol 执行文稿，以将组特异残基映射入（标记在）MutH 蛋白结构中。将 MutH 结构载入 RasMol 程序，在“RasMol Command Line”用“Script”命令执行从 GeneDoc 输出的文稿，以观察组特异残基，辨别位于假设为 DNA 结合位点的候选残基。

（3）下载限制性内切核酸酶与 DNA 底物（或产物）形成复合物的现有序列。运行 Swiss PDB Viewer，以 3 个催化残基（如 MutH 的 D70、E77 和 K79）的主链原子为种子，将限制性内切核酸酶结构与目标 MutH 结构进行匹配。用“Improve fit”功能增强拟合效果。将重叠结构的坐标输出到电子制表程序。计算目标蛋白（如 MutH）的任意原子到重叠结构中碱基的距离，此碱基对应于目标蛋白中的目标碱基。对所有重叠结构重复这些步骤。计算平均距离，并辨认对感兴趣碱基距离最近的残基。将结果与组特异残基分析进行比较，为定点突变选择有希望的残基。

（4）MutH 变体的克隆。采用如 Kirsch 和 Joly 所修改过的 QuikChange 操作规程（Stratagen），以质粒 pMQ402 为模板，以及两个寡聚脱氧核苷酸用于突变以适合于产生长度在 50～500 bp 的 PCR 产物（见注意事项 2）。

（5）将细菌质粒以标准分子生物学方法转化 XL－1 Blue MRF 细胞。将细胞平铺在含有氨苄西林的 LB 培养板上，于 37 ℃过夜培养。挑选单一菌落接种于 500 mL 含氨苄西林的 LB 培养液中，于 37 ℃生长。在 600 nm 处光密度值为 0.8 时，于 28 ℃，用 0.2%（m/V）阿拉伯糖（终浓度）诱导 2.5 小时（见注意事项 3）。3 000 转/分钟离心细胞 10 分钟。去除上清液后，将细胞重悬于 10 mL 结合缓冲液中，用超声波破碎器破碎（输出级 5，占空率 50%，超声破碎 5 次，每次 1 分钟，冷却间隔 1 分钟）。

（6）在冷冻离心机上，以 30 000 转/分钟离心细胞碎片 30 分钟。于 4 ℃，将上清液与 1.5 mL Ni－NTA 琼脂糖浆柔和地混合 30 分钟。将 Ni－NTA 琼脂糖移至一空柱中，用 20 mL 清洗缓冲液冲洗该柱，用 0.5 mL 洗脱缓冲液洗脱。没有必要切除 His 标记，因其不干扰内切酶活性（见注意事项 4）。以在 280 nm 处的光密度值作为判断依据，收集合并得到的每份蛋白质。于 4 ℃，用 500 mL 透析缓冲液透析样品最少 2 小时。透析过程中，换缓冲液 2 次，用 500 mL 透析缓冲液 G 透析样品。在透析缓冲液中以 1∶10 比例稀释样品，测量 280 nm 处的光吸收，以便用理论消光系数计算 MutH 的摩尔浓度，于－20 ℃保存蛋白质。

（7）采用含有未甲基化、半甲基化或过甲基化单 d（GATC）位点（通过某种方法合成的寡核苷酸或 PCR 产物）的 DNA 底物对 MutH 变体进行 MutH 和 Cut 分析。采用不同的荧光染料（分别为 FAM 和 TET）标记底物上下两端的链，以检测每个链上的断裂。

（8）将浓度为 10 nmol/L 的 DNA 底物置于 10 μL 的测试缓冲液中，该缓冲液包含 500 nmol/L的 MutL 和 10～500 nmol/L 的 MutH，并将反应混合物在 37 ℃下孵育（见注意事项 5）。

（9）以适当的时间间隔（10 秒到 30 分钟）分装反应混合物，每份含 25 fmol PCR 产物，与 12 μL 模板抑制剂（Perkin-Elmer）和 0.5 μL GeneScan－500 TAMRA 标准品充分混合。加热到 95 ℃ 2 分钟，并立即在冰上冷却。在配有 47 cm（内径 50 μm）毛细管（其中含有加了8 mol/L尿素的 POP－4 聚合物）的基因分析仪上分析样品。

（10）用 5 秒将样品注入毛细管，在电压 15 000 V 及 60 ℃下，使用 1×基因分析缓冲液外加 1 mmol/L EDTA 作为电极缓冲液，用 30 分钟完成电泳过程。记录切割的和未切割的荧光标记的 DNA 数量测定切割速度。

【注意事项】

（1）通过在“Edit Sequence Groups”对话框中设置“Group Cons Level”和“PCRMaximum Level”可以改变图像阴影。

（2）一个引物会导致期望的突变，而另一个为反义引物，以生成所谓“megaprimer”用于定点突变（据 QuikChange 操作规程）。

（3）诱导的时间和温度取决于被研究的体系。

（4）某些情况下，需切除 His 标记。此处使用的构建允许在 His 标记和 N 端甲硫氨酸间的凝血酶位点做切割。

（5）与使用 MutL 活化不同的另一种方法，是加入 10%（V/V）二甲基亚砜，可使 MutH 内切酶活性激活到 10 倍。

【实验结果】

产生针对半甲基化 d（GATC）位点具有不同特异性的 MutH 变体。

XV. Protein Engineering of Site-Specific Endonuclease

【Objectives】

(1) Learn to change the various steps involved in the site-specific endonuclease specific design project.

(2) Learn the method for analyzing the structural changes of protein.

(3) Learn to identify candidate amino acid residues responsible for sensing the methylation status.

【Principle】

When the conversion mismatch repair nuclease MutH is stimulated by MetS and MutL, it cuts a gap at the hemimethylation site d (GATC), which is converted to cut permethylated, hemimethylated and unmethylated DNA variant.

【Software, apparatus, materials and reagents】

1. Software

ClustalW (http://www.ebi.ac.uk/clustalw/), ClustalX (http://bips.u-strasbg.fr/fr/Documentation/ClustalX/), GeneDoc (http://www.psc.edu/biomed/genedoc/), RasMol (http://www.openrasmol.org/), Swiss PDB Viewer (http://www.expasy.org/spdbv/).

2. Apparatus

Ultrasonic breaker, freezing centrifuge, empty column, GeneScan-500 TAMRA DNA Molecular Quality Standard and ABI PRISM 310 Genetic Analyzer.

3. Materials

pMQ402, PCR primers, pfu polymerase, deoxyribonucleoside triphosphate, arabinose, benzyl fluoride (PMSF), Ni-NTA agarose, *E. coli* strain XL-1 Blue MRF, *E. coli* MutL protein, fluorescent group-labeled oligonucleotides or fluorescent-labeled PCR products, template DNA, phosphorylated primer light-group-labeled primers, dam methylase, λ-exonuclease, exonuclease, P-spin, pfu DNA polymerase and 47 cm capillary tube filled with POP- 4 polymer with 8 mol/L urea (inner diameter: 50 μm).

4. Reagents

(1) LB medium (per liter): add 10 g tryptone for bacteria, 5 g yeast extract for bacteria and 5 g sodium chloride in distilled water. Adjust pH to 7. 2 –7. 5 with sodium hydroxide and sterilize.

(2) Binding buffer: 20 mmol/L Tris-HCl, 5 mmol/L imidazole, 1 mol/L sodium chloride and 1 mmol/L PMSF (pH 7. 9) .

(3) Washing buffer: 20 mmol/L Tris-HCl, 60 mmol/L imidazole, 1 mol/L sodium chloride and 1 mmol/L PMSF (pH 7. 9) .

(4) Elution buffer: 20 mmol/L Tris-HCl, 200 mmol/L imidazole, 1 mol/L sodium chloride and 1 mmol/L PMSF (pH 7. 9) .

(5) Dialysis buffer: 10 mmol/L HEPES potassium hydroxide, 100 mmol/L potassium chloride, 1 mmol/L EDTA and 1 mmol/L dithiothreitol (pH 7. 9) .

(6) Dialysis buffer G: 10 mmol/L HEPES potassium hydroxide, 100 mmol/L potassium chloride, 1 mmol/L EDTA, 1 mmol/L dithiothreitol and 50% glycerol (pH 7. 9) .

(7) Test buffer: 10 mmol/L Tris-HCl (pH 7. 5), 10 mmol/L magnesium chloride, 0. 75 mmol/L adenosine triphosphate and 0. 1 mg/mL bovine serum albumin.

(8) 1 × gene analysis buffer with 1 mmol/L EDTA.

(9) Template inhibitor.

【Procedures】

(1) From the basic local alignment search tool (BLAST) page of the National Center for Biotechnology Information (http: //www. ncbi. nlm. nikh go/BLAST) to obtain the MutH protein and related restriction endonucleases sequence. More sequences are obtained from the PSI BLAST server. Use the PAM250 matrix and the ClustalX program to align the sequences, and apply the GeneDoc program to analyze the aligned sequences. How to use the program can be obtained by using the "Help" function in the program. The coordinates of the structure are obtained from the PDB database (http: //www. resb. org/pdb/). Programs RasMol and Swiss PDB Viewer are used for structure observation.

(2) Enter the alignment results into the GeneDoc program, and use the "Group" function to group sequences (for example, one group contains MutH and the other contains restriction endonucleases) to identify group-specific residues. Use the "RasMol Script Dialog" function in GeneDoc to generate RasMol execution scripts to map group-specific residues to (marked in) the MutH protein structure. Load the MutH structure into the RasMol program, and use the "Script"

command in the "RasMol Command Line" to execute the manuscript output from GeneDoc to observe the group-specific residues. Identify candidate residues that are hypothesized to be DNA binding sites.

(3) Download the existing sequence of restriction endonuclease and DNA substrate (or product) to form the existing sequence of the complex. Run the Swiss PDB Viewer and use the main chain atoms of 3 catalytic residues (such as D70, E77 and K79 of MutH) as seeds to match the restriction endonuclease structure with the target MutH structure. Use the "Improve fit" function to enhance the fit effect. Output the coordinates of the overlapping structure to the spreadsheet program. Calculate the distance from any atom of the target protein (such as MutH) to the base in the overlapping structure. This base corresponds to the target base in the target protein. Repeat these steps for all overlapping structures. Calculate the average distance and identify the residue closest to the base of interest. Compare the results with group-specific residue analysis. Choose promising residues for site-directed mutagenesis.

(4) Cloning of MutH variants. Use the modified QuikChange operating procedure (Stratagen) as described by Kirsch and Joly, use plasmid pMQ402 as a template, and two oligodeoxynucleotides for mutation to be suitable for generating lengths in 50 – 500 bp PCR product (See Notes 2).

(5) XL – 1 Blue MRF cells are transformed with bacterial plasmids by standard molecular biological methods. The cells are spread on LB culture plate containing ampicillin and cultured overnight at 37 ℃. First, a single colony is grown in 500 mL LB medium containing ampicillin at 37 ℃. 0.2% (m/V) arabinose (final concentration) is used to induce its growth at 28 ℃ for 2.5 h when the optical density value is 0.8 at 600 nm (see Notes 3). The cells are then centrifuged at 3 000 rpm for 10 mins. Remove the supernatant and the cells are resuspended in 10 mL of binding buffer and broken by ultrasonic breaker (output level 5, duty cycle 50%, ultrasonic crushing 5 times, 1 min each time, 1 min interval for cooling).

(6) In a freezing centrifuge, centrifuge the cell debris at 30 000 rpm for 30 mins, at 4 ℃, gently mix the supernatant with 1.5 mL Ni-NTA agar syrup for 30 mins, and transfer the Ni-NTA agarose to an empty column. Rinse the column with 20 mL washing buffer. Elute with 0.5 mL elution buffer. It is not necessary to remove the His mark, because it does not interfere with endonuclease activity (see Notes 4). Take the optical density value at 280 nm as the judgment basis, and collect each protein obtained by merging. Dialysis the sample with 500 mL dialysis buffer at 4 ℃ for a minimum of 2 h. Change the buffer twice and dialyze the sample with 500 mL of dialysis buffer G. Dilute the sample at a ratio of 1 : 10 in the dialysis buffer, measure the light absorption at 280 nm, in order to calculate the molar concentration of MutH with the theoretical extinction coefficient, and store the protein at – 20 ℃.

(7) Use DNA substrates containing unmethylated, hemimethylated or permethylated single d (GATC) sites (oligonucleotides synthesized by some means, or PCR products) to carry out MutH and Cut analysis of MutH variants. Different fluorescent dyes (FAM and TET respectively) are used to label the chains at the upper and lower ends of the substrate in order to detect the cleavage on each chain.

(8) The DNA substrate at a concentration of 10 nmol/L is in 10 μL of test buffer containing MutL at a concentration of 500 nmol/L and MutH at a concentration of 10 – 500 nmol/L, and the reaction mixture is incubated at 37 ℃ (see Notes 5).

(9) Dispense the reaction mixture at appropriate time intervals (10 s to 30 mins), each containing 25 fmol PCR product, and mix well with 12 μL template inhibitor (Perkin-Elmer) and 0. 5 μL GeneScan-500 TAMRA Standard. Heat to 95 ℃ for 2 mins, and immediately cool on ice. Analyze the sample on ABI PRISM 310 Genetic Analyzer equipped with 47 cm (inner diameter 50 μm) capillary (containing POP- 4 polymer with 8 mol/L urea).

(10) In 5 s, inject the sample into the capillary, using a voltage of 15 000 V at 60 ℃, using 1 × gene analysis buffer plus 1 mmol/L EDTA as the electrode buffer, and use 30 mins to complete the electrophoresis process. Record the number of cleaved and uncut fluorescently labeled DNA to determine the cutting speed.

【Notes】

(1) The shadow of the image can be changed by setting "Group Cons Level" and "PCRMaximum Level" in the "Edit Sequence Groups" dialog box.

(2) One primer will cause the desired mutation, while the other is an antisense primer to generate a so-called "megaprimer" for site-directed mutation (according to QuikChange operating procedures).

(3) The induction time and temperature depend on the system being studied.

(4) In some cases, the His mark needs to be removed. The construct used here allows cleavage at the thrombin site between the His mark and the N-terminal methionine.

(5) Another method different from MutL activation is to add 10% (V/V) dimethyl sulfoxide to activate MutH endonuclease activity up to 10 times.

【Result】

MutH variants with different specificties for the hemimethylation d (GATC) site are produced.

十六、M13噬菌体衣壳蛋白改造在改良噬菌体展示技术中的应用

【实验目的】

掌握改进蛋白质展示水平的噬菌体展示库的设计方法。

【实验原理】

噬菌体展示是一项改造结合靶分子多肽的强大技术。蛋白质与噬菌体衣壳蛋白形成融合蛋白，可以在噬菌体表面表达，而相应的编码基因则包裹于噬菌体颗粒内。包裹的DNA通过突变扩增构建噬菌体展示多肽库，为筛选各种变异体做准备。通过固定化靶分子体外结合实验，高亲和力的蛋白质可以从蛋白质展示库中筛选出来。随后，选定蛋白质的序列可以通过DNA测序推断出来。靶蛋白可以通过构建噬菌粒载体与主要衣壳蛋白［蛋白质-8（P8）］或次要衣壳蛋白［蛋白质-3（P3）］形成的融合蛋白在M13噬菌体表面以低拷贝方式展示，这需要在野生型（wt）P8和P3辅助噬菌体的帮助下进行。

【实验材料】

1. P8突变库的构建

（1）制备含有尿嘧啶的单链DNA模板。

①2YT培养基：在容量瓶中加入10克酵母抽提物、16克菌用胰蛋白胨和5克氯化钠，加水到1 L，用氢氧化钠调节pH到7.0，高压灭菌。

②2YT/carb/cmp培养基：在2YT培养基中加入50 μg/mL羧苄青霉素和5 μg/mL氯霉素。

③2YT/carb/kan/uridine培养基：在2YT培养基中加入50 μg/mL羧苄青霉素、25 μg/mL卡那霉素和0.25 μg/mL尿嘧啶。

④羧苄青霉素：5 mg/mL羧苄青霉素水溶液，超滤灭菌。

⑤氯霉素：50 mg/mL氯霉素乙醇溶液。

⑥大肠杆菌CJ236。

⑦卡那霉素：5 mg/mL卡那霉素水溶液，超滤灭菌。

⑧生理磷酸缓冲液（PBS）：在容量瓶中加入137 mmol/L氯化钠、3 mmol/L氯化钾、8 mmol/L磷酸氢二钠以及1.5 mmol/L磷酸二氢钾，用盐酸（HCl）调节pH到7.2，高压灭菌。

⑨聚乙二醇（polyethylene glycol，PEG）/NaCl：在容量瓶中加入20% PEG－8 000（m/V）和2.5 mol/L氯化钠，高压灭菌。

⑩QIAprep SpinM13试剂盒。

⑪M13KO7辅助噬菌体。

⑫Tris-acetate－EDTA（TAE）缓冲液：在容量瓶中加入40 mmol/L Tris-acetate和1 mmol/L EDTA，调节pH到8.0，高压灭菌。

⑬TAE琼脂糖胶：取TAE缓冲液，加入1%琼脂糖（m/V）以及1∶5 000（V/V）10%溴化乙锭（EB），于烧杯中混合，加热使琼脂糖溶化，冷却后备用。

⑭尿嘧啶：25 mg/mL尿嘧啶水溶液，超滤灭菌。

（2）体外合成异质共价闭环双链DNA。

①100 mmol/L二硫苏糖醇（DTT）。

②25 mmol/L dNTP：含有dATP，dCTP，dGTP和dTTP各25 mmol/L的溶液。

③10 mmol/L ATP。

④10×Tris－Mg（TM）缓冲液：在容量瓶中加入0.1 mol/L $MgCl_2$以及0.5 mol/L Tris－HCl，pH 7.5。

⑤QIAquick Gel Extraction试剂盒。

⑥T_4聚核苷酸激酶。

⑦TA DNA连接酶。

⑧TAE琼脂糖凝胶。

⑨超纯水。

（3）大肠杆菌电穿孔及噬菌体扩增。

①2YT/carb培养基：2YT及50 μg/mL羧苄青霉素。

②羧苄青霉素（carbenicillin，carb）。

③电转化感受态*E. coli* SS320。

④ 0.2厘米间隙电穿孔样品杯。

⑤ECM－600电穿孔仪（BTX）。

⑥卡那霉素（kanamycin，kan）。

⑦Luria-Bertani（LB）/carb培养皿：LB琼脂及50 μg/mL羧苄青霉素。

⑧PBS，见第（1）项下⑧条。

⑨PEG/NaCl，见第（1）项下⑨条。

⑩SOC培养基：在容量瓶中加入5克菌用酵母抽提物、20克菌用胰蛋白胨和0.5克氯化钠及0.2克氯化钾，加水到大约1 L，用氢氧化钠调节pH到7.0，高压灭菌。然后加入

5 mL 2.0 mol/L 高压灭菌过的氯化镁溶液及 20 mL 超滤灭菌过的 1.0 mol/L 葡萄糖溶液，并将最终体积调至 1L。

⑪M13KO7 辅助噬菌体，见第（1）项⑪条。

（4）用电穿孔法制备 *E. coli* SS320 感受态细胞。

①1.0 mmol/L HEPES（pH 7.4）：将 4.0 mL 1.0 mol/L 的 HEPES（pH 7.4）加入 4.0 L的超纯水，超滤灭菌。

②10%（V/V）的超纯甘油：100 mL 超纯甘油加入 900 mL 超纯水中，超滤灭菌。

③2YT/tet 培养基：2YT 及 5 μg/mL 四环素。

④磁性搅拌棒（约 5 厘米长），放入乙醇中浸泡。

⑤Superbroth/tet 培养基：在锥形瓶中加入 24 克菌用酵母抽提物、12 克细菌用胰蛋白胨和 5 mL 甘油，加水到 900 mL，高压灭菌。然后，加入 100 mL 高压灭菌的 0.17 mol/L 磷酸二氢钾和 0.72 mol/L 的磷酸氢二钾混合液，最后加入 5 μg/mL 四环素。

⑥四环素（tetracycline）：5 mg/mL 四环素水溶液，超滤灭菌。

⑦超纯甘油。

⑧超纯水。

2. 选择和分析增加融合蛋白展示水平的 P8 变异体

（1）从 hGH－P8 库中选择噬菌体。

① 0.2% 牛血清白蛋白（BSA）的 PBS 溶液。

②100 mmol/L HCl。

③1.0 mol/L Tris 碱。

④ 96 孔 maxisorp 免疫板。

⑤大肠杆菌 *E. coli* XL－1 Blue 菌株。

⑥PBS－T 缓冲液：PBS 及 0.05% Tween－20。

⑦PBS－T－BSA 缓冲液：PBS，0.05% Tween－20 以及 0.2% BSA。

⑧2YT/carb /kan 培养基：2YT，50 μg/mL 羧苄青霉素以及 25 μg/mL 卡那霉素。

（2）噬菌体酶联免疫筛选法测定 hGH 的展示水平。

1.0 mol/L H_3PO_4，2YT/carb/kan 培养基，3，3'，5'，5－四甲基联苯胺（TMB）/H_2O_2过氧化物酶底物，96 孔 maxisorp 免疫板，羧苄青霉素（carb），大肠杆菌 *E. coli* XL－1 Blue 菌株，辣根过氧化物酶（Horseradish peroxidase）/抗 M13 抗体络合物，卡那霉素，LB/tet 培养皿（LB 琼脂及 5 μg/mL 四环素），M13KO7 helper 噬菌体，PBS，PBS－T 缓冲液，PBS－T－BSA 缓冲液，PEG/NaCl，四环素。

【实验步骤】

1. 库的设计

在噬菌粒表达系统中 P8 的 N 端部分极度耐受突变，其中有些突变能增加异源融合蛋白的展示水平。P8 的 N 端只有 6 个野生型残基侧链（*Ala*7，*Ala*5，*Ala*10，*Phe*11，*Leu*14 和 *Ala*18），对 P8 有效包装到噬菌体外壳是必需的，这些侧链形成一个紧密的疏水抗原决定簇，在噬菌体组装过程中起到关键作用。环绕这一决定簇的残基突变能增加包装效率，从而增加异源蛋白的展示水平。基于这些研究，7 个位点（*Pro*6，*Lys*8，*Asn*12，*Ser*13，*Gln*15，*Ala*16 和 *Ser*17）的突变被认为最可能改进蛋白质展示水平。蛋白质内 7 个位点的完全随机化能导致接近 10^9 的独特氨基酸组合，这一多样性能够被这里所描述的制备库的约 10^{10} 的多样性所覆盖。

2. 库的构建

（1）纯化 dU - ssDNA 模板。

①从新鲜的 LB/抗菌素板上挑出一个含有某噬菌粒的 *E. coli* CJ236（或者其他 dut^-/ung^-）菌株的单克隆，将其放入 1 mL 2YT 加有 M13KO7 helper 噬菌体（10^{10} pfu/mL）和相应抗菌素的培养基中，以保持宿主 F’附加体和噬菌粒。例如，2YT/carb/cmp 培养基中的羧苄青霉素用来选择带有 β - 内酰胺酶基因的噬菌粒，而氯霉素用来选择 CJ236F’附加体。37 ℃及 200 转/分钟下摇动 2 小时并且加入卡那霉素（25 μg/mL）来选择被 M13KO7 共转染的带有卡那霉素抗性基因的克隆。在 37 ℃及 200 转/分钟下摇动 6 小时后，转移到 30 mL 的 2YT/carb/kan/uridine 培养基培养。再在 37 ℃及 200 转/分钟下摇动过夜。

②在 4 ℃ 27 000 转/分钟下离心 10 分钟，转移上清液到含有 1/5 体积 PEG/NaCl 的新试管中，室温孵育 5 分钟。在 4 ℃ 12 000 转/分钟下离心 10 分钟，移走上清液，在 4 000 转/分钟下短暂离心，吸出残余上清液。

③在 0. 5 mL 的 PBS 中重新悬浮起噬菌体沉淀小团，转移到一个新的 EP 离心管中。在桌式小离心机中用 14 000 转/分钟离心 5 分钟，转移上清液到新的 EP 离心管中。

④加入 7. 0 μL MP 缓冲液，混匀。室温孵育至少 2 分钟。

⑤在 2 mL 小离心试管的 QIAprep spin column（离心层析柱）中加入上述样品。在桌式小离心机中 12 000 转/分钟离心 30 秒，遗弃流出液。噬菌体颗粒仍然结合在层析柱介质中。

⑥在柱中加入 0. 7 mL MLB 缓冲液，12 000 转/分钟离心 30 秒，弃流出液。

⑦在柱中再加入 0. 7 mL MLB 缓冲液。室温孵育至少 1 分钟，12 000 转/分钟离心 30 秒，弃流出液。噬菌体 DNA 与壳蛋白分离，仍然吸附在层析柱介质中。

⑧加入 0. 7 mL PE 缓冲液，12 000 转/分钟离心 30 秒，弃流出液。

⑨重复步骤⑧，去除残余的蛋白质及盐。

⑩12 000 转/分钟离心 30 秒，将小层析柱转移到一个新的 1. 5 mL EP 离心试管中。

⑪加入 100 μL 的 EB 缓冲液（10 mmol/L Tris - HCl，pH 8.5）到层析柱膜的中心部位。室温孵育 10 分钟，然后 12 000 转/分钟离心 30 秒，保存流出液，其中包含纯化的dU - ssDNA。

⑫分析上述 DNA，用 1.0 μL DNA 溶液进行 TAE 琼脂糖凝胶电泳。结果 DNA 应该是明显的单一条带，但是也经常可以观察到一些较低电泳迁移率的弱带。

⑬利用 260 nm 的吸收值（A_{260} = 1.0 相应于 33 ng/μL 单链 DNA）测定 DNA 浓度。典型 DNA 浓度范围应该为 200 ~ 500 ng/μL。

（2）体外合成异源双链 CCC - dsDNA。

①用 T_4 多聚核苷酸激酶磷酸化寡核苷酸。

a. 在 1.5 mL EP 离心管中加入 0.6 μg 突变寡核苷酸、2.0 μL 的 10 × TM 缓冲液、2.0 μL 10 mmol/L ATP 和 1.0 μL 100 mmol/L DTT 混合，加水到总体积为 20 μL。

b. 往上述体系中加入 20 U T_4 多聚核苷酸激酶，37 ℃孵育 1 小时。

②寡核苷酸和模板一起退火。

a. 在 20 μL 磷酸化反应混合物中加入 20 μg 的 dU - ssDNA 模板、25 μL 的 10 × TM 缓冲液，加水到总体积为 250 μL，假设寡核苷酸与模板长度比是 1 : 100，上述 DNA 的量给出寡核苷酸与模板摩尔比为 3 : 1。

b. 90 ℃孵育 3 分钟，50 ℃孵育 3 分钟，20 ℃孵育 5 分钟。

③合成 CCC - dsDNA。

a. 在已经退火的寡核苷酸与模板混合物中加入 10 μL 10 mmol/L ATP，10 μL 25 mmol/L 的 dNTP，15 μL 100 mmol/L DTT，30 U T_4 DNA 连接酶以及 30 U 的 T_7 DNA 聚合酶。

b. 20 ℃孵育过夜。

c. 利用 Qiagen QIAquick DNA 纯化试剂盒，对上述 DNA 进行亲和纯化及脱盐。加入 1.0 mL QG 缓冲液后混合。

d. 将上述样品放入两个置于 2 mL 离心试管中的 QIAquick spin 离心层析柱。在桌式小离心机中用 14 000 转/分钟离心 1 分钟，去除流过液。

e. 在每个柱子中加入 750 μL PE 缓冲液，14 000 转/分钟离心 1 分钟，去除流过液，再 13 000 转/分钟离心 1 分钟，将层析柱置于新的 1.5 mL EP 小离心管。

f. 在层析柱膜的中心部位加入 35 μL 超纯水，室温孵育 2 分钟。

g. 14 000 转/分钟离心 1 分钟，用水洗提出 DNA 两次，收集并混合两个管中的洗提液。

h. 同时于 dU - ssDNA 模板电泳 1.0 μL 上述洗提出的反应产物，用含有溴化乙锭（EB）的 TAE 琼脂糖凝胶电泳来检测 DNA。

（3）大肠杆菌电穿孔和噬菌体扩增。

①将上述纯化好的 DNA（约 20 μg）及 0.2 厘米间隙的电转槽置于冰上冷却。在冰上

融化 350 μL 分装好的感受态 *E. coli* SS320 细胞。将感受态细胞加入 DNA 中且用移液枪吸放数次混合（避免产生气泡）。

②转移混合物于电转槽中进行电穿孔转化。电穿孔转化要根据仪器出产商提供的说明书进行操作。

③立即在电穿孔转化后的电转槽中加入 1 mL SOC 培养基以营救这些细胞并且转移培养基到一个 250 mL 锥形瓶中。用 1 mL SOC 培养基冲洗电转槽两次，再向瓶中加入 SOC 培养基到终体积为 25 mL，在 200 转/分钟摇床 37 ℃孵育 20 分钟。

④要确定所建库的多样性，可以在 LB/carb 培养皿中系列稀释涂板来选择适当的噬菌粒（如带有 β－内酰胺酶基因的噬菌粒 pS1607）。

⑤加入 M13KOT helper 噬菌体（4×10^{10} pfu/mL），在 200 转/分钟摇床 37 ℃孵育 10 分钟。

⑥转移培养液于含有 500 mL 2YT 培养基的 2 L 锥形瓶中，加入合适的抗生素进行噬菌粒选择（如 2YT/carb 培养基）。

⑦在 200 转/分钟摇床 37℃孵育 1 小时且加入 25 μg/mL 卡那霉素，然后在 200 转/分钟摇动下 37 ℃孵育过夜。

⑧将上述培养液在 4 ℃及 16 000 转/分钟下离心 10 分钟。转移上清液到含有 1/5 体积 PEG/NaCl 的新试管中以沉淀噬菌体。室温孵育 5 分钟。

⑨在 4 ℃及 16 000 转/分钟下离心 10 分钟，去除上清液。再短暂离心一下，用移液管吸走上清液残余。重新悬浮噬菌体小团于 1/20 体积的 PBS 中。

⑩在 4 ℃及 27 000 转/分钟下离心 5 分钟以去除不溶性杂质。转移上清液于一个干净试管。

⑪利用分光光度计估计噬菌体浓度（268 nm 的光密度［OD_{268}］＝1.0 时，溶液中噬菌体浓度约为 5×10^{12} 噬菌体/mL）。

（4）用电穿孔法制备 *E. coil* SS320 感受态细胞。

①在 1 mL 2YT/tet 培养基中接种一个生长于新鲜 LB/tet 培养皿的单克隆 *E. coli* SS320 菌株。在 200 转/分钟摇床 37℃孵育 6～8 小时。

②转移培养液于有 500 mL 2YT/tet 培养基的 2L 锥形瓶中，在 200 转/分钟摇动下 37℃孵育过夜。

③用 5 mL 上述过夜培养液接种 6 个 2 L 锥形瓶，每瓶含有 900 mL 的 Superbroth/tet 培养基。在 200 转/分钟摇动下 37 ℃孵育到 OD_{550} 约为 0.8。

④在冰上冷却 3 个上述锥形瓶 5 分钟，不时摇动。下列步骤⑤～⑫应该在冷室中及冰上进行，所用一切溶液及仪器应该预冷。

⑤采用 Sorvall GS－3 台式离心机，在 4 ℃及 5 000 转/分钟下离心 10 分钟，去除上清液，然后加入从其他 3 个瓶中余下的培养液（所有液体需要预冷）。重复上述离心及去除

上清液步骤。

⑥在离心瓶中加入 1.0 mmol/L HEPES（pH 7.4），同时加入灭菌的磁铁搅拌棒来帮助重悬离心沉淀的细胞。摇动使沉淀脱离管壁，并且在适中的速度下磁力搅拌以完全重悬细胞沉淀。

⑦采用 Sorvall GS－3 台式离心机，在 4 ℃及 5 500 转/分钟下离心 10 分钟，去除上清液，小心管中的磁铁搅拌棒。从转子中取出离心管时要小心保持离心管的位置，以避免干扰管底的细胞沉淀。

⑧在离心瓶中加入 1.0 mmol/L HEPES，pH 7.4，重悬细胞沉淀，重复步骤⑥和⑦中的重悬及离心过程。去除上清液。

⑨重悬每管细胞沉淀于 150 mL 的 10% 超纯甘油中，不要混合各离心管。

⑩在 4 ℃及 5 000 转/分钟下离心 15 分钟，去除上清液，取出磁铁搅拌棒，用一个移液管小心吸出残余的上清液。

⑪在一个离心管中加入 3.0 mL 的 10% 超纯甘油，用移液枪小心抽吸重悬细胞沉淀。转移悬浮好的细胞到另一个离心管，重复上述过程直到所有细胞沉淀都得到很好的悬浮。

⑫350 μL 的感受态细胞在液氮中快速被冰冻分装并且储存于－70 ℃。

3. 选择和分析能提高融合蛋白展示水平的 P8 突异体

（1）从 hGH－P8 库中选择噬菌体。

①用靶蛋白包被 96 孔 Maxisorp 板。

a. 用 100 μL 5 μg/mL 的靶蛋白（如 hGHbp）溶液包被 96 孔 Maxisorp 板中的 8 孔，4 ℃孵育过夜。直接倾倒 96 孔板于水池，以去除孔中溶液。

b. 在每孔中加入 200 μL 0.2% 溶于 PBS 的 BSA 溶液以阻止其他蛋白质对 Maxisorp 板的非特异性结合，室温下摇动 1 小时。

②噬菌体库的选择。

a. 在上述每孔中加入一定量的置于 PBS－T－BSA 缓冲液中的噬菌体库（约 10^{12} phages/mL），在室温下摇动 2 小时。然后用 PBS－T 缓冲液洗 96 孔 Maxisorp 板 8 次。

b. 在每孔中加入 100 μL 的 100 mmol/L HCl 以洗脱结合的噬菌体。在室温下强力摇动 5 分钟。

c. 将所有洗脱液收集到一起，加入 1/5 体积的 1.0 mol/L Tris－碱中和。

③增殖噬菌体以备后用。

a. 将上述洗脱出的噬菌体混合液加入 10 倍体积的 XL－1 Blue 细胞中（OD_{550} =0.5～1.0）。

b. 37 ℃下 200 转/分钟摇动孵育 20 分钟，取出 10 μL 留备测量滴度。

c. 加入 M13KO7 helper 噬菌体，在 200 转/分钟摇床 37 ℃孵育 45 分钟。

d. 转移培养液于 100 mL 的 2YT/carb/kan 培养基中，在 200 转/分钟，37 ℃下摇动过夜。

e. 利用 PEG/NaCl 沉淀法分离噬菌体。

f. 重复上述筛选过程 5 次，每轮只用一半的洗脱噬菌体。

（2）噬菌体酶联免疫法测定 hGH 展示水平。

①从新鲜的 LB/tet 板上挑出一个含有特定噬菌粒的 *E. coli* XL－1 Blue 菌株的单克隆，放入 1 mL 加有 M13KO7 helper 噬菌体（10^{10} pfu/mL）和 50 μg/mL 羧苄青霉素（以保持噬菌体）及 5 μg/mL 四环素（以保持 F' 附加体）的 2YT 培养基中。在 200 转/分钟，37 ℃下摇动 2 小时，再加入 25 μg/mL 的卡那霉素来选择共转染了 M13KO7 的克隆。再在 200 转/分钟，37 ℃下摇动 6 小时，转移培养液于 30 mL 2YT/carb/kan 培养基中。在 200 转/分钟，3 ℃下摇动过夜。

②在 4 ℃及 27 000 转/分钟下离心 10 分钟，转移上清液于一个含有 1/5 体积 PEG/NaCl 的干净试管中，室温孵育 5 分钟。在 4 ℃及 12 000 转/分钟下离心 10 分钟，去除上清液。在 4 000 转/分钟下短暂离心一下，再用移液管吸走上清液残余。

③重悬噬菌体沉淀小团于 0.5 mL 的 PBS－T－BSA 缓冲液中，将其转移到一个 1.5 mL 的 EP 离心管中。上清液在桌式离心机中用 14 000 转/分钟离心 5 分钟，再转移到一个 1.5 mL的 EP 离心管中。

④利用分光光度计法估计噬菌体浓度［$\varepsilon = 1.2 \times 10^{8}$/（pfu/mL）］。

⑤用 PBS－T－BSA 缓冲液准备系列 5 倍稀释的噬菌体储液。

⑥转移 100 μL 的噬菌体溶液到有 hGHbp 包被且 BSA 封闭的 96 孔 Maxisorp 板，温和摇动孵育 1 小时。

⑦去除噬菌体溶液，用 PBS－T 缓冲液洗板 8 次。

⑧加入 100 μL 的辣根过氧化物酶（horseradish peroxidase）/抗 M13 抗体络合物（用 PBS－T－BSA 缓冲液稀释 3 000 倍），温和摇动孵育 30 分钟。

⑨用 PBS－T 缓冲液洗 8 次，再用 PBS 洗 2 次。

⑩用 100 μL 的 TMB 底物溶液显影 96 孔免疫板；用 100 μL 的 1.0 mol/L H_3PO_4溶液终止反应，然后在 96 孔板读板仪中读出 450 nm 的光吸收值。

【注意事项】

磷酸化的寡核苷酸应该立即使用。环状 DNA 的电泳迁移率取决于盐浓度、pH 和 EB 的存在与否。DNA 必须在直接加有 EB 的 TAE/琼脂糖凝胶上进行电泳，切忌将 EB 加到电泳缓冲液中。冷冻能使有些展示蛋白变性，使之变得不稳定，不利于选择。一般而言，最好尽快使用展示库。

【实验结果】

筛选出能提高融合蛋白展示水平的 P8 噬菌体变异体。

XVI. Application of M13 Phage Capsid Protein Modification in Improved Phage Display Technology

【Objective】

Master the designing method of phage display library to improve protein display level.

【Principle】

Phage display is a powerful technology for engineering peptides that bind to target molecules. The protein forms a fusion protein with the phage capsid protein, which can be expressed on the surface of the phage, and the corresponding coding gene is wrapped in the phage particle. The wrapped DNA is used to construct a phage display peptide library through mutation and amplification, also to prepare for the screening of various variants. Through *in vitro* binding experiments of immobilized target molecules, high-affinity protein can be screened from the protein display library. Subsequently, the sequence of the selected protein can be inferred by DNA sequencing. The target protein can be displayed on the surface of M13 phage in a low-copy manner by constructing a fusion protein formed by the phagemid vector and the major capsid protein [Protein-8 (P8)] or the minor capsid protein [Protein-3 (P3)]. This needs to be done with the help of wild-type (wt) P8 and P3 helper phages.

【Materials】

1. Construction of P8 mutation library

(1) Preparation of single-stranded DNA template containing uracil.

①2YT medium: take 10 g yeast extract, 16 g tryptone for bacteria and 5 g sodium chloride into a volumetric flask and add water to 1 L. Adjust the pH to 7.0 with sodium hydroxide, then autoclave it.

②2YT/carb/cmp medium: add 50 μg/mL carbenicillin and 5 μg/mL chloramphenicol into the 2YT medium.

③2YT/carb/kan/uridine medium: add 50 μg/mL carbenicillin, 25 μg/mL kanamycin and 0. 25 μg/mL uracil into the 2YT medium.

④Carbenicillin: prepare 5 mg/mL carbenicillin aqueous solution and sterilize by ultrafiltration.

⑤Chloramphenicol: 50 mg/mL chloramphenicol ethanol solution.

⑥*E. coli* CJ236.

⑦Kanamycin: prepare 5 mg/mL kanamycin aqueous solution and sterilize by ultrafiltration.

⑧Physiological phosphate buffer (PBS): add 137 mmol/L sodium chloride, 3 mmol/L potassium chloride, 8 mmol/L disodium hydrogen phosphate and 1. 5 mmol/L potassium dihydrogen phosphate in a volumetric flask and adjust pH to 7. 2 with hydrochloric acid (HCl) . Autoclave it.

⑨Polyethylene glycol (PEG) /NaCl: prepare 20% PEG-8 000 (m/V) and 2. 5 mol/L sodium chloride in a volumetric flask and autoclave it.

⑩QIAprep SpinM13 kit.

⑪M13KO7 helper phage.

⑫Tris-acetate-EDTA (TAE) buffer: prepare 40 mmol/L Tris-acetate and 1 mmol/L EDTA in a volumetric flask and adjust pH to 8. 0. Autoclave it.

⑬TAE agarose gel: take TAE buffer, add 1% agarose (m/V) and 1 : 5 000 (V/V) 10% ethidium bromide (EB) in a beaker and mix it. Heat to dissolve agarose and cool for use.

⑭Uracil: prepare 25 mg/mL uracil aqueous solution and sterilize by ultrafiltration.

(2) *In vitro* synthesis of heterogeneous covalent closed-loop double-stranded DNA.

①100 mmol/L dithiothreitol (DTT).

②25 mmol/L dNTP: a solution containing dATP, dCTP, dGTP and dTTP each 25 mmol/L.

③10 mmol/L ATP.

④10 × Tris-Mg (TM) buffer: prepare 0. 1 mol/L $MgCl_2$ and 0. 5 mol/L Tris-HCl in a volumetric flask, and adjust pH to 7. 5.

⑤QIAquick Gel Extraction kit.

⑥T_4 polynucleotide kinase.

⑦TA DNA ligase.

⑧TAE agarose gel.

⑨Ultra pure water.

(3) *E. coli* electroporation and phage amplification.

①2YT/carb medium: add 50 μg/mL carbenicillin into 2YT.

②Carbenicillin (carb) .

③Electrotransformation competent *E. coli* SS320.

④0. 2 cm gap electroporation sample cup.

⑤ECM-600 electroporator（BTX）.

⑥Kanamycin（kan）.

⑦Luria-Bertani（LB）/carb Petri dish：add 50 μg/mL carbenicillin in LB agar medium.

⑧PBS，see ⑧ in（1）.

⑨PEG/NaCl，see ⑨ in（1）.

⑩SOC medium：take 5 g yeast extract for bacteria，20 g tryptone for bacteria，0. 5 g sodium chloride and 0. 2 g potassium chloride in a volumetric flask；add water to about 1L，adjust the pH with sodium hydroxide to 7. 0. Autoclave it. Then add 5 mL 2. 0 mol/L autoclaved magnesium chloride solution and 20 mL ultrafiltration sterilized 1. 0 mol/L glucose solution into the medium. Adjust the final volume to 1 L.

⑪M13KO7 helper phage. see ⑪ in（1）.

（4）*E. coli* SS320 competent cells are prepared by electroporation.

①1. 0 mmol/L HEPES（pH 7. 4）：Add 4. 0 mL 1. 0 mol/L HEPES（pH 7. 4）to 4. 0 L ultrapure water and sterilize by ultrafiltration.

②10%（V/V）ultrapure glycerin：add 100 mL ultrapure glycerin into 900 mL ultrapure water and sterilize by ultrafiltration.

③2YT/tet medium：add 5 μg/mL tetracycline into 2YT.

④Magnetic stirring rod（about 5 cm long）. Soak it in ethanol.

⑤Superbroth/tet medium：take 24 g yeast extract for bacteria，12 g tryptone for bacteria and 5 mL glycerol in an Erlenmeyer flask and add water to 900 mL. Autoclave it. Then add 100 mL of autoclaved 0. 17 mol/L potassium dihydrogen phosphate and 0. 72 mol/L dipotassium hydrogen phosphate mixture，and finally add 5 μg/mL tetracycline.

⑥Tetracycline：prepare 5 mg/mL tetracycline aqueous solution and sterilize it by ultrafiltration.

⑦Ultrapure glycerin.

⑧Ultrapure water.

2. Selection and analysis of P8 variants that increase the display level of fusion proteins

（1）Select phage from the hGH-P8 library.

①0. 2% bovine serum albumin（BSA）PBS solution.

②100 mmol/L HCl.

③1. 0 mol/L Tris base.

④96 well maxisorp immune plate.

⑤*E. coli* XL-1 Blue strain.

⑥PBS-T buffer: add 0.05% Tween-20 in PBS.

⑦PBS-T-BSA buffer: add 0.05% Tween-20 and 0.2% BSA in PBS.

⑧2YT/carb/kan medium: add 50 μg/mL carbenicillin and 25 μg/mL kanamycin in the 2YT medium.

(2) Use phage-enzyme linked immunoscreening method to determine the display level of hGH.

1.0 mol/L H_3PO_4, 2YT/carb/kan medium, 3, 3', 5', 5 – tetramethylbenzidine (TMB) / H_2O_2 peroxidase substrate, 96 well maxisorp immune plate, Carbenicillin (carb), *E. coli* XL-1 Blue strain, Horseradish peroxidase/anti-M13 antibody complex, Kanamycin, LB/tet Petri dish (add 5 μg/mL tetracycline in LB agar medium), M13KO7 helper phage, PBS, PBS-T buffer, PBS-T-BSA buffer, PEG/NaCl, tetracycline.

【Procedures】

1. Design library

In the phagemid expression system, the N-terminal part of P8 is extremely tolerant of mutations, some of which can increase the display level of heterologous fusion proteins. There are only 6 wild-type residue side chains (*Ala*7, *Ala*5, *Ala*10, *Phe*11, *Leu*14 and *Ala*18) at the N-terminus of P8, which are necessary for the effective packaging of P8 into the phage coat. These side chains form a tight hydrophobic epitope, which plays a key role in the phage assembly process. Mutations in residues surrounding this determinant can increase packaging efficiency, thereby increasing the display level of heterologous protein. Mutations at 7 sites (*Pro*6, *Lys*8, *Asn*12, *Ser*13, *Gln*15, *Ala*16 and *Ser*17) are most likely to improve the display level of protein. The complete randomization of 7 sites in the protein can result in a unique combination of amino acids close to 10^9. This diversity can be covered by the diversity of about 10^{10} in the preparation library described here.

2. Construct library

(1) Purified dU-ssDNA template.

①Pick a single clone of *E. coli* CJ236 (or other dut^-/ung^-) strain containing a certain phagemid from a fresh LB/antibiotic plate and put it into 1 mL 2YT plus M13KO7 helper phage (10^{10} pfu/mL) and corresponding antibiotic medium to maintain the host F' episome and phagemid. For example, 2YT/carb/cmp medium contains carbenicillin to select phagemids with β – lactamase gene, while chloramphenicol is used to select CJ236F' episome. Shake it at 37 ℃ and 200 rpm for 2 h and add kanamycin (25 μg/mL) to select clones co-transfected by M13KO7 with kanamycin resistance gene. After shaking at 37 ℃ and 200 rpm for 6 h, transfer the culture to 30 mL of 2YT/carb/kan/uridine medium. Then shake it overnight at 37 ℃ and 200 rpm.

②Centrifuge for 10 mins at 4 ℃ 27 000 rpm, transfer the supernatant to a new test tube containing 1/5 volume of PEG/NaCl, and incubate it at room temperature for 5 mins. Centrifuge for 10 mins at 4 ℃ at 12 000 rpm, remove the supernatant, centrifuge briefly at 4 000 rpm, and aspirate the remaining supernatant.

③Resuspend the phage pellet in 0.5 mL of PBS and transfer to a new EP centrifuge tube. Centrifuge at 14 000 rpm for 5 mins in a small table-type centrifuge, and transfer the supernatant to a new EP centrifuge tube.

④Add 7.0 μL MP buffer and mix them well. Incubate it at room temperature for at least 2 mins.

⑤Add the above samples to the QIAprep spin column in a small 2 mL centrifuge tube. Centrifuge at 12 000 rpm for 30 s in a small table-type centrifuge and discard the effluent. The phage particles are still bound in the chromatography column medium.

⑥Add 0.7 mL of MLB buffer to the column, centrifuge at 12 000 rpm for 30 s, and discard the effluent.

⑦Add 0.7 mL of MLB buffer to the column. Incubate at room temperature for at least 1 min, centrifuge at 12 000 rpm for 30 s. Discard the effluent. The phage DNA is separated from the shell protein and still adsorbed in the chromatography column medium.

⑧Add 0.7 mL of PE buffer, centrifuge at 12 000 rpm for 30 s, and discard the effluent.

⑨Repeat step ⑧ to remove residual protein and salt.

⑩Centrifuge at 12 000 rpm for 30 s and transfer the small column to a new 1.5 mL EP centrifuge tube.

⑪Add 100 μL of EB buffer (10 mmol/L Tris-HCl, pH 8.5) to the center of the column membrane. Incubate it at room temperature for 10 mins, and then centrifuge at 12 000 rpm for 30 s to save the effluent, which contains purified dU-ssDNA.

⑫Analyze the above-mentioned DNA and perform TAE agarose gel electrophoresis with 1.0 μL DNA solution. As a result, the DNA should be an obvious single band, but some weak bands with lower electrophoretic mobility can often be observed.

⑬The DNA concentration is determined using the absorbance value of 260 nm ($A_{260}=1.0$ corresponds to 33 ng/μL single-stranded DNA). The typical DNA concentration range should be 200 – 500 ng/μL.

(2) *In vitro* synthesis of heteroduplex CCC-dsDNA.

①Phosphorylate oligonucleotides with T_4 polynucleotide kinase.

a. In a 1. 5 mL EP centrifuge tube, add 0. 6 μg of mutant oligonucleotide, 2. 0 μL of 10 × TM buffer, 2. 0 μL of 10 mmol/L ATP and 1. 0 μL of 100 mmol/L DTT. Mix them well and add water to the total volume of 20 μL.

b. Add 20 U T_4 polynucleotide kinase to the above system. Incubate it at 37 ℃ for 1 h.

②Oligonucleotide and template are annealed together.

a. Add 20 μg of dU-ssDNA template, 25 μL of 10 × TM buffer to 20 μL of phosphorylation reaction mixture, add water to a total volume of 250 μL, assuming that the length ratio of oligonucleotide to template is 1 : 100, the amount of DNA above gives a molar ratio of oligonucleotide to template of 3 : 1.

b. Incubate it at 90 ℃ for 3 mins, 50 ℃ for 3 mins, 20 ℃ for 5 mins.

③Synthesize CCC-dsDNA.

a. Add 10 μL of 10 mmol/L ATP, 10 μL of 25 mmol/L dNTP, 15 μL of 100 mmol/L DTT, 30 U of T_4 DNA ligase and 30 U of T_7 DNA polymerase to the annealed oligonucleotide and template mixture.

b. Incubate it overnight at 20 ℃.

c. Use Qiagen QIAquick DNA purification kit to perform affinity purification and desalting of the above DNA. Add 1. 0 mL of QG buffer and mix them.

d. Put the above samples into two QIAquick spin columns in 2 mL centrifuge tubes. Centrifuge at 14 000 rpm for 1 min in a small table-type centrifuge to remove the flow-through.

e. Add 750 μL of PE buffer to each column, centrifuge at 14 000 rpm for 1 min, remove the flow-through, then centrifuge at 13 000 rpm for 1 minute and put the column in a new 1. 5 mL EP microcentrifuge tube.

f. Add 35 μL of ultrapure water into the center of chromatography column membrane. Incubate it at room temperature for 2 mins.

g. Centrifuge at 14 000 rpm for 1 min, wash the extracted DNA twice. Collect the eluate and mix them well.

h. At the same time, perform electrophoresis to 1. 0 μL of the reaction products produced by the above washing with dU-ssDNA template, detect DNA by TAE agarose gel electrophoresis containing ethidium bromide (EB).

(3) Electroporation and phage amplification of *E. coli*.

①Cool the purified DNA (about 20 μg) and a 0. 2 cm gap electrorotating bath on ice. 350 μL of *E. coli* SS320 competent cells are thawed on ice. The competent cells are added to DNA and sucked several times with a pipette gun to mix them (avoid bubbles).

②The mixture is transferred to an electrorotating bath for electroporation conversion. The operation of electroporation conversion should follow the instruction manual of the instrument manufacturer.

③Immediately after electroporation conversion, 1 mL of SOC medium is added to the electroporation bath to rescue these cells, and the medium is transferred into a 250 mL conical flask. The electroporation bath is washed twice with 1 mL SOC medium. Then, add the SOC medium to the bottle until the final volume is 25 mL, and incubate at 37 ℃ for 20 mins at 200 rpm shaker.

④To determine the diversity of the constructed library, a series of dilution plates can be used in LB/carb dishes to select the appropriate phagemid (such as pS1607 with β-lactamase gene).

⑤Add M13KOT helper phage (4×10^{10} pfu/mL), incubate it in a shaker with 200 rpm at 37 ℃ for 10 mins.

⑥Transfer the culture medium into a 2 L conical flask containing 500 mL 2YT medium, and add appropriate antibiotics for phagemid selection (such as 2YT/carb medium).

⑦Incubate in 200 rpm shaker at 37 ℃ for 1 h and add 25 μg/mL kanamycin. Then incubate it overnight at 37 ℃ under 200 rpm shaking.

⑧Centrifuge the above culture medium at 4 ℃ and 16 000 rpm for 10 mins. The supernatant is transferred to a new test tube containing 1/5 volume of PEG/NaCl to precipitate the phage. Incubate at room temperature for 5 mins.

⑨The supernatant is removed by centrifugation at 4 ℃ and 16 000 rpm for 10 mins. After a short centrifugation, pipette the supernatant residue. The phage clusters are resuspended in 1/20 volume of PBS.

⑩The insoluble impurities are removed by centrifugation at 4 ℃ and 27 000 rpm for 5 mins. Transfer the supernatant to a clean tube.

⑪Estimate the concentration of phage by spectrophotometer (when the optical density of 268 nm [OD_{268}] = 1.0, the concentration of phage in solution is about 5×10^{12} phages/mL).

(4) *E. coli* SS320 competent cells are prepared by electroporation.

①A monoclonal *E. coli* SS320 strain cultured in fresh LB/tet Petri dish is inoculated in 1 mL 2YT/tet medium. The cells are incubated at 37 ℃ for 6 – 8 h on a 200 rpm shaker.

②The transferred medium is incubated overnight at 37 ℃ under 200 rpm shaking in a 2 L conical flask with 500 mL 2YT /tet medium.

③Six 2 L conical flasks containing 900 mL of Superbroth/tet medium are inoculated with 5 mL of the above 3 overnight medium. The OD_{550} is about 0.8 when incubated at 37 ℃ under 200 rpm shaking.

④Cool the above 3 conical flasks on ice for 5 mins and shake it from time to time. The following steps ⑤ to ⑫ should be carried out in a cold chamber and on ice. All solutions and instruments used should be precooled.

⑤Centrifugation at 4 ℃ and 5 000 rpm for 10 mins is carried out in the Sorvall GS-3 rotor. The supernatant is removed, and then the remaining culture medium is added from the other three bottles (all liquids need to be precooled). Repeat the above centrifugation and supernatant removal steps.

⑥1. 0 mmol/L HEPES (pH 7. 4) and sterilized magnetic stirring rod are added to the centrifugation bottle to help resuspend the centrifuged cells. The precipitate is shaken off the tube wall, and the cell precipitate is completely suspended by magnetic stirring at a moderate speed.

⑦Centrifugate at 4 ℃ and 5 500 rpm in the rotor of Sorvall GS-3 for 10 mins, remove the supernatant, and be careful of the magnetic stirring rod in the tube. When removing the centrifuge tube from the rotor, the position of the centrifuge tube should be carefully kept to avoid interfering with the cell precipitation at the bottom of the tube.

⑧Add 1. 0 mmol/L HEPES into the centrifuge flask, and then resuspend the cells for precipitation at pH 7. 4. Repeat the resuspension and centrifugation in steps ⑥ and ⑦. Remove the supernatant.

⑨The cells in each tube are resuspended and precipitated in 150 mL of 10% ultrapure glycerin, and do not mix the centrifugation tubes.

⑩The supernatant is removed by centrifugation at 4 ℃ and 5 000 rpm for 15 mins. The magnetic stirring rod is taken out. The residual supernatant is carefully sucked out with a pipette.

⑪Add 3. 0 mL of 10% ultrapure glycerin into a centrifuge tube, and carefully suck the suspended cell precipitate with pipette. Transfer the suspended cells to another centrifuge tube and repeat the above process until all cells are suspended well.

⑫350 μL competent cells are rapidly frozen in liquid nitrogen and stored at −70 ℃.

3. Selection and analysis of P8 variants that can improve the display level of fusion proteins

(1) Selection of phages from hGH-P8 library.

①Coating 96 well Maxisorp plate with target protein.

a. Eight wells in 96 well Maxisorp plates are coated with 100 μL of 5 μg/mL target protein (such as hGHbp) solution and incubated overnight at 4 ℃. The 96 well plate should be directly poured into the pool to remove the solution in the hole.

b. 200 μL of 0. 2% BSA solution in PBS is added to each well to prevent the non-specific binding of other proteins to the Maxisorp plate, and the plate should be shaken at room temperature for 1 h.

②Selection of phage library.

a. A certain amount of phage library (about 10^{12} phages/mL) in PBS-T-BSA buffer is added into each well and shaken at room temperature for 2 h. Then the 96 well Maxisorp plate is washed with PBS-T buffer for 8 times.

b. 100 μL of 100 mmol/L HCl is added to each well to elute the bound phage. Shake strongly at room temperature for 5 mins.

c. All eluates are collected and neutralized with 1/5 volume of 1.0 mol/L Tris-base.

③Proliferation of phage for later use.

a. The washed-out phage mixture is added into 10 times the volume of XL-1 Blue cells (OD_{550} = 0.5 – 1.0).

b. Shake and incubate it at 37 ℃ for 20 mins at 200 rpm, then take out 10 μL to reserve for titer measurement.

c. M13KO7 helper phage is added and incubated at 37 ℃ for 45 mins at 200 rpm shaking table.

d. Transfer the culture to 100 mL 2YT/carb/kan medium. Shake it overnight at 37 ℃ at 200 rpm.

e. The phage is isolated by PEG/NaCl precipitation method.

f. The above screening process is repeated 5 times, and only half of the eluted phages are used in each round.

(2) Use phage-enzyme linked immunosorbent assay to determine hGH display level.

①A single clone of *E. coli* XL-1 Blue strain containing specific phagemid is selected from the fresh LB/tet plate and inoculated in 1 mL 2YT medium supplemented with M13KO7 helper phage (10^{10} pfu/mL), 50 μg/mL carboxybenzylpenicillin (to maintain phage) and 5 μg/mL tetracycline (to maintain F' appendage). At 200 rpm, shake at 37 ℃ for 2 h and then add 25 μg/mL kanamycin to select the co-transfected M13KO7 clone. After shaking at 200 rpm and 37 ℃ for 6 h, the culture is transferred to 30 mL 2YT/carb/kan medium. Shake it overnight at 200 rpm and 3 ℃.

②Centrifugate at 4 ℃ and 27 000 rpm for 10 mins. The supernatant is transferred to a clean test tube containing 1/5 volume of PEG/NaCl and incubated at room temperature for 5 mins. The supernatant is removed by centrifugation at 4 ℃ and 12 000 rpm for 10 mins. After a short centrifugation at 4 000 rpm, the residual supernatant is removed by pipette.

③The phage precipitates are suspended in 0.5 mL PBS-T-BSA buffer and transferred to a 1.5 mL EP centrifuge tube. The supernatant is centrifuged with 14 000 rpm for 5 mins in a table-type centrifuge, and then transferred to a 1.5 mL EP centrifuge tube.

④Estimation of phage concentration by spectrophotometer [$\varepsilon = 1.2 \times 10^8/(\text{pfu/mL})$].

⑤PBS-T-BSA buffer is used to prepare serial 5-fold dilutions of phage storage.

⑥Transfer 100 μL of phage solution to 96 well Maxisorp plate coated with hGHbp and blocked by BSA. Incubate it with mild shaking for 1 h.

⑦The phage solution is removed and the plates are washed with PBS-T buffer for 8 times.

⑧Add 100 μL of horseradish peroxidase/anti-M13 antibody complex (diluted 3 000 times with PBS-T-BSA buffer). Incubate it for 30 mins with mild shaking.

⑨Wash 8 times with PBS-T buffer and 2 times with PBS.

⑩ 96 well immune plates are developed with 100 μL TMB substrate solution. The reaction is terminated with 100 μL 1.0 mol/L H_3PO_4 solution, and then the optical absorption value at 450 nm is read out in a 96 well plate reader.

【Notes】

Phosphorylated oligonucleotides should be used immediately. The electrophoretic mobility of cyclic DNA depends on the concentration of salt, pH and the presence of EB. DNA must be electrophoresed on TAE/agarose gel directly added with EB, and EB should not be added to the electrophoretic buffer. Freezing can denaturate some display proteins and make them unstable, which is not conducive to selection. In general, it's best to use the library as soon as possible.

【Result】

The P8 phage variant is screened to improve the display level of fusion proteins.

十七、蛋白酶产生菌的筛选

【实验目的】

了解并掌握蛋白酶产生菌的分离纯化方法。

【实验原理】

在固体培养基中渗入溶解性差、可被特定菌利用的营养成分，如可溶性淀粉、奶粉或纤维素等，会造成浑浊、不透明的培养基背景。接种培养一定时间后，有些菌落周围就会形成水解圈，其大小与菌落利用此物质的能力有关，从而达到简便、高效筛选的目的。例如，在脱脂奶粉培养基中可根据水解圈的有无和大小来初步筛选产蛋白酶的菌株。

为了获得较好的实验结果，通常采用双层平板培养基进行水解圈观察实验。该方法的优点主要有：①加入底层培养基可弥补培养皿底部不平的缺陷；②可使所有的水解圈都位于近乎同一平面上，大小一致、边缘清晰且无重叠现象；③因上层培养基中琼脂含量减半，可形成形态较大、特征较明显的水解圈以便于观察和测量。

【实验仪器、材料与试剂】

1. 仪器

振荡器、恒温培养箱、超净工作台、显微镜等。

2. 材料

市售豆腐乳、培养皿、试管、接种环、无菌移液管、无菌玻璃涂棒。

3. 试剂

马铃薯葡萄糖培养基（PDA 培养基），筛选培养基（脱脂奶粉 1.5%、琼脂 1.0%，pH 6.0，灭菌备用）。

【实验步骤】

1. 制备豆腐乳菌悬液

称取 10 g 豆腐乳至 250 mL 三角烧瓶中，加入 90 mL 无菌水和适量玻璃珠，振荡 15 min，充分混匀。于超净工作台内用无菌移液管取 1 mL 上述菌悬液，加入盛有 9 mL 无菌水的试管中充分混匀，制备成10^{-2}稀释度的菌悬液。然后采用同样操作方式，从上述试管中取1 mL加至另一个盛有9 mL无菌水的试管中，混合均匀，制备成10^{-3}稀释度的菌悬液。以此类推制备10^{-4}、10^{-5}、10^{-6}不同稀释倍数的菌悬液备用。

2. 双层平板的制备

将事先配制并经灭菌处理的筛选培养基和PDA培养基加热溶化。于超净工作台内先用PDA培养基在培养皿上倒一层培养基，待凝固后，倒入筛选培养基。一般下层PDA培养基倒入15 mL，上层筛选培养基倒入10 mL。

3. 涂布

用无菌移液管分别从10^{-3}、10^{-4}、10^{-5}、10^{-6}四管稀释液中各取0.1 mL加入平板中，用无菌玻璃涂棒涂布均匀，静置30分钟，使菌液吸附进培养基。在每个平板底部用记号笔做好标记以便区分，重复3次。

4. 蛋白酶产生菌的分离

将筛选培养基平板倒置于28 ℃恒温培养箱中培养2～3天，观察水解圈的有无与大小。从水解圈直径与菌落直径的比值较大的菌落上，分别挑取少许细胞接种到PDA培养基的斜面上，并进行编号。待斜面上长出菌苔后，镜检确定是否为单一微生物。若发现有杂菌，需采用平板划线法对菌株进行纯化直到能实现纯培养。

【实验结果】

将相应的测量数据填入表17－1。

表17－1　蛋白酶产生菌记录表

	菌株编号					
	菌株1号	菌株2号	菌株3号	菌株4号	菌株5号	……
菌落直径/mm						
水解圈直径/mm						
水解圈直径/菌落直径						

【思考题】

（1）水解圈法筛选产蛋白酶菌株的原理是什么？

（2）为什么要以水解圈直径与菌落直径的比值作为筛选指标？

XVII. Screening of Protease-Producing Bacteria

【Objective】

Understand and master the isolation and purification method of protease-producing bacteria.

【Principle】

In the solid medium, some nutrients, such as soluble starch, milk powder or cellulose, which are poorly soluble and can be used by specific bacteria, are infiltrated, resulting in a turbid and opaque background in medium. After inoculating and culturing for a certain period of time, hydrolysis circle will be formed around some colonies, the size of which is related to the ability of colonies to use this material, so as to achieve the purpose of simple and efficient screening. For example, the screening of the protease strains depends on the presence or absence and size of the hydrolysis circle in the non-fat milk powder medium.

In order to obtain better experimental results, the observation experiment of hydrolysis circle is usually carried out using double-layer media plate. The advantages of this method are: ①the bottom layer of medium can make up for the defect of uneven bottom of the dish; ②all the hydrolysis circles can be located in the same plane, with the same size, clear edges and no overlap; ③because the agar content in the upper layer of medium is halved, larger and more distinctive hydrolysis circles can be formed to facilitate observation and measurement.

【Apparatus, materials and reagents】

1. Apparatus

Shaker, constant temperature incubator, ultra-clean worktable, microscope, etc.

2. Materials

Commercially available sufu (fermented bean curd), Petri dishes, test tubes, inoculation rings, sterile pipettes and sterile glass applicator rod.

3. Reagents

Potato dextrose medium (PDA medium) and screening medium (non-fat milk powder

1.5%, agar 1.0%, pH 6.0, sterilize and ready to use).

【procedures】

1. Preparation of microbial suspension from sufu

Weigh 10 g of sufu into a 250 mL triangular flask, add 90 mL of sterile water and appropriate amount of glass beads. Shake the flask for 15 mins, mix the solution thoroughly. Take 1 mL of the above suspension into a tube with a sterile pipette in the ultra-clean worktable, supplement 9 mL of sterile water and mix thoroughly to prepare a 10^{-2} dilution of bacterial suspension. Then follow the same steps to take 1 mL from the above test tube to another test tube with 9 mL sterile water, mix them well and prepare 10^{-3} dilution of bacterial suspension. To prepare 10^{-4}, 10^{-5}, 10^{-6} different dilutions of bacterial suspension by following the same operation for backup.

2. Preparation of bilayer plates

The prepared and sterilized screening medium and PDA medium are heated and dissolved. In the ultra-clean worktable, pour a layer of medium first with PDA medium on Petri dish. After solidification, pour the screening medium onto the PDA medium. Generally, the lower layer of PDA medium is about 15 mL, and the upper layer of screening medium is about 10 mL.

3. Coating

Use a sterile pipette to take 0.1 mL of each of the four dilutions (10^{-3}, 10^{-4}, 10^{-5}, 10^{-6}) onto the plate. Use a sterile glass applicator rod to spread the solution evenly on the surface of plates and let stand for 30 mins, so that the bacterial solution will be adsorbed into the medium. Mark the bottom of each plate with a marker to distinguish, and repeat 3 times.

4. Isolation of protease-producing bacteria

Place the screening medium plate upside down in a constant temperature incubator at 28 ℃ for 2 – 3 days, and observe the presence and size of hydrolysis circles. From the colonies with larger ratio of hydrolysis circle diameter to colony diameter, a small number of cells are inoculated onto the slant of PDA medium and numbered. After the moss grows on the slant, microscopic examination is performed to determine whether it is a single microorganism. If there are different strains observed, we need to use the plate scribing method to purify the strain until we get a pure culture.

【Result】

Fill in the corresponding experimental data into Table XVII – 1.

Table XVII-1 Protease-producing bacteria

	Strain number					
	Strain No. 1	Strain No. 2	Strain No. 3	Strain No. 4	Strain No. 5	...
Colony diameter/mm						
Hydrolysis circle diameter/mm						
Hydrolysis circle diameter/colony diameter						

【Questions】

(1) What is the principle of screening protease-producing strains by the hydrolysis circle method?

(2) Why should the ratio of hydrolysis circle diameter to colony diameter be used as the screening index?

十八、细菌的紫外线诱变

【实验目的】

(1) 理解并掌握紫外线诱变的原理和方法。

(2) 学习单孢子悬液制备方法及影印培养技术。

【实验原理】

DNA 可强烈地吸收紫外线，紫外线会造成 DNA 链的断裂、DNA 分子内和分子间的交联、核酸与蛋白质的交联、胞嘧啶和尿嘧啶的水合作用及碱基二聚体的形成等 DNA 分子的改变，从而诱发突变或使微生物死亡。在上述可引起细胞突变的因素中，形成碱基二聚体是主要原因。碱基二聚体可以在同一条链相邻的碱基之间产生，也可以在两条链的碱基之间形成，嘧啶比嘌呤对紫外线敏感得多，因此，交联的二聚体多为嘧啶二聚体（尤其是胸腺嘧啶二聚体）。嘧啶二聚体会引起 DNA 复制错误，正常碱基无法配对，造成复制错义或缺失。过量的紫外线照射会造成菌体丢失大片段的 DNA，或使交联的 DNA 无法打开，不能进行复制或转录，从而引起菌体死亡。

紫外线对生物的效应具有积累作用，就是说只要紫外线处理的总时间相等，分次处理与一次性处理的效果类似。紫外线造成的 DNA 损伤是可以得到及时修复的，如果将受紫外线照射后的细胞立即暴露在可见光下，菌体的突变率和死亡率都会有所下降，这一现象称为光复活作用。光复活作用主要是由受可见光激活的光复活酶执行的，光复活酶会识别胸腺嘧啶二聚体，并与之结合形成复合物，此时光复活酶没有活性。可见光光能可以激活光复活酶，使之将二聚体打开从而使 DNA 复原。因一般的微生物细胞内都具有光复活酶，所以，在进行紫外线诱变育种时，应尽量在避光或红光条件下操作。

鉴于白光对紫外线诱变后的菌株有较为强烈的修复作用，可以在诱变过程中，采用致死剂量的紫外线和白光交替处理待诱变的微生物，使其突变率及突变幅度进一步提高。此外，紫外线诱变时，正突变一般多出现在偏低的剂量中，而负突变则多出现在偏高剂量中，经多次诱变而产量得到提高的菌株更容易出现负突变。

【实验仪器、材料与试剂】

1．仪器

生化培养箱、紫外灯、磁力搅拌器、影印模具、移液枪等。

2．材料

生产上使用的卡那霉素链霉菌（*Streptomyces kanamyceticus*）菌株、牙签、锥形瓶等。

3．试剂

（1）卡那霉素链霉菌培养基。

（2）含卡那霉素的分离用培养基平板。

卡那霉素母液制备：用去离子水配制卡那霉素终浓度为400 mg/mL，用无菌过滤器过滤除菌备用。

制备含卡那霉素的卡那霉素链霉菌分离培养基平板：制备含卡那霉素300 μg/mL、500 μg/mL、900 μg/mL、1 200 μg/mL、1 500 μg/mL、2 500 μg/mL、3 500 μg/mL及8 000 μg/mL的分离用培养基平板，同时制作含10 000 μg/mL，50 000 μg/mL，100 000 μg/mL卡那霉素（可直接加入粉剂）的分离培养基平板。

（3）卡那霉素链霉菌发酵培养基。

【实验步骤】

1．出发菌株性能测定

（1）出发菌株卡那霉素生产能力测定：将出发菌株卡那霉素链霉菌斜面种子接种于液体培养基中，27 ℃、200 转/分钟摇床培养140 小时，按照《中华人民共和国药典》2005版提供的卡那霉素测定方法测定其含量。

（2）出发菌株耐自身代谢终产物水平测定：用10 mL生理盐水将出发菌株斜面上的孢子洗下，倒入装有小玻璃珠的锥形瓶中，充分摇动，使孢子分散，之后用脱脂棉漏斗过滤，获得单孢子悬浮液。各取0.1 mL孢子悬浮液，分别涂布于含有300 μg/mL、600 μg/mL、900 μg/mL、1 200 μg/mL、1 500 μg/mL 、2 500 μg/mL、3 500 μg/mL卡那霉素的分离培养基平板上（每个浓度重复三皿），27 ℃培养6～9 天，观察平板上菌落生长情况。如果在某一浓度的平板上形成菌落，而下一个较高浓度的平板不生成菌落，则该浓度就是选定的初筛的耐药药物浓度。

2．诱变处理

（1）为防止光照回复突变的发生，整个诱变、诱变后涂布及培养过程，均需避光进行。

（2）取10 mL单孢子悬液于ϕ90 培养皿中（带无菌磁力搅拌棒），将皿放置于诱变箱中的磁力搅拌器上，皿放置在紫外灯正下方30厘米处。

(3) 开启紫外灯，预热20分钟后，开启磁力搅拌器，打开皿盖，分别照射15秒、30秒、45秒、60秒、75秒、90秒。

(4) 取对照（不经紫外线诱变）和不同时间诱变处理的孢子悬液各1.0 mL，进行适当稀释，各取0.1 mL涂布于分离用培养基平板上，27 ℃培养2~3天，计算细胞存活率。

3. 产物耐受性突变株的筛选

(1) 按照出发菌株的初始卡那霉素耐受能力（例如，在前述实验中获知，该卡那霉素生产株的卡那霉素耐受浓度低于1 200 μg/mL，即在1 200 μg/mL平板上不生长，则选择1 200 μg/mL作为筛选分离平板的卡那霉素浓度）制作筛选分离平板，取处理后的孢子悬液和对照孢子悬液各0.1 mL，涂布于筛选分离平板上，27 ℃培养6~7天，统计生长的抗性菌落，挑入斜面保存。

(2) 制作含有卡那霉素为1 500 μg/mL、2 500 μg/mL、3 500 μg/mL的分离培养基平板，将上述挑出的菌株斜面孢子各取1环在平板中划线，27 ℃培养6~8天。从中挑出10~15株对卡那霉素耐受性较高的菌株接种于斜面保存。

(3) 将上述筛出的抗性菌株划线接种于制备好的含有卡那霉素为10 000 μg/mL、50 000 μg/mL、100 000 μg/mL的分离培养基平板上，27 ℃培养6~8天，观察结果，选择对卡那霉素耐受性较高的菌株接种于斜面保存。

(4) 将上述能够耐受较高浓度卡那霉素（10 000 μg/mL、50 000 μg/mL、100 000 μg/mL）的突变株接种至发酵培养基进行发酵，测定其卡那霉素的效价，从中筛选出耐高浓度卡那霉素的高产突变株。

【实验结果】

(1) 根据实验结果计算突变率及突变量，分析育种过程中存在哪些问题。

(2) 记录不同存活率条件下获得的突变株的数量，讨论如何提高突变效率。

【思考题】

(1) 试述紫外线诱变时需要注意哪些问题。

(2) 为什么紫外线诱变需要在有搅拌的培养皿中进行？还可以采用什么方式进行诱变？

XVIII. UV Mutagenesis of Bacteria

【Objectives】

(1) Understand and master the principle and method of UV mutagenesis.

(2) Learn the preparation of single spore suspension and photo culture technique.

【Principle】

DNA can strongly absorb ultraviolet light. Ultraviolet light will cause DNA strand breaks, DNA molecules and intermolecular cross-linking, nucleic acid and protein cross-linking, cytosine and uracil hydration, the formation of base dimers and other changes in DNA molecules, thereby inducing mutations or the death of microorganisms. Among the above-mentioned factors that can cause cell mutation, the formation of base dimers is the main cause. Base dimers can be formed between adjacent bases of the same strand or between bases of two strands. Pyrimidines are much more sensitive to ultraviolet light than purines; therefore, most cross-linked dimers are pyrimidine dimers (especially thymine dimers) . Pyrimidine dimers cause DNA replication errors, where normal bases fail to pair, resulting in replication missense or deletion. Excessive UV exposure can cause bacterium to lose large segments of DNA or prevent cross-linked DNA from opening and replication or transcription from taking place, resulting in death of bacterium.

The effects of UV on organisms are cumulative, meaning that multiple treatment has similar effects to one-time treatment as long as the total time of UV treatment is equal. The DNA damage caused by UV can be repaired in time, and if the UV-irradiated cells are exposed to visible light immediately, the mutation rate and mortality rate of the organism will be reduced, causing a phenomenon known as photoreactivation. Photoreactivation is mainly performed by photoreactivation enzyme activated by visible light. Photoreactivation enzyme recognizes thymine dimer and binds to it to form a complex and hence photoreactivation enzyme is inactive. Visible light activates photoreactivation enzyme, which opens the dimer to recover DNA. Since microbial cells in general have photoreactivation enzyme, UV mutagenesis breeding should be carried out under dark or red light condition as much as possible.

Since white light has a strong repairing effect on the UV mutagenized strains, the mutation rate and mutation magnitude can be further increased by alternating lethal doses of UV and white light to the microorganisms to be mutagenized during the mutagenesis process. In addition, the positive mutations generally appear in the low dose of UV mutagenesis, while the negative mutations appear in the high dose, and the strains with improved yield by multiple mutagenesis are more likely to have negative mutations.

【Apparatus, materials and reagents】

1. Apparatus

Biochemical incubator, ultraviolet lamp, magnetic stirrer, photocopy molds and pipettes, etc.

2. Materials

Streptomyces kanamyceticus strain used in the production, toothpicks, Erlenmeyer flasks.

3. Reagents

(1) *Streptomyces kanamyceticus* medium.

(2) Medium plate containing kanamycin for isolation.

Preparation of kanamycin mother liquor: Prepare a kanamycin solution (400 mg/mL) with deionized water, filter with sterile filter and set aside.

Preparation of *Streptomyces kanamyceticus* isolation medium plates containing kanamycin: Prepare medium plates containing 300 μg/mL, 500 μg/mL, 900 μg/mL, 1 200 μg/mL, 1 500 μg/mL, 2 500 μg/mL, 3 500 μg/mL and 8 000 μg/mL of kanamycin for isolation, and also prepare medium plates containing 10 000 μg/mL, 50 000 μg/mL, 100 000 μg/mL kanamycin (can be added directly to the powder) .

(3) *Streptomyces kanamyceticus* fermentation medium.

【Procedures】

1. Determine the original strain performance

(1) Determination of kanamycin production capacity of the original strain: The slanted seeds of *Streptomyces kanamyceticus* are inoculated in liquid medium, and culture for 140 h at 27 ℃ and 200 rpm. The content of kanamycin is determined according to the determination method of kanamycin provided by *Pharmacopoeia of the People's Republic of China* (2005 Edition) .

(2) Determination of the resistance level of the original strain to its own end products of metabolism: Wash the spores on the inclined surface of the original strain with 10 mL normal saline, pour them into an Erlenmeyer flask with small glass beads. Shake them fully to disperse the spores,

and then filter them with a absorbent cotton funnel to obtain a single spore suspension. Take 0. 1 mL of spore suspension, respectively, and spread it on the separation medium plate containing 300 μg/mL, 600 μg/mL, 900 μg/mL, 1 200 μg/mL, 1 500 μg/mL, 2 500 μg/mL and 3 500 μg/mL kanamycin (repeat three dishes for each concentration) . Culture it for 6 – 9 d at 27 ℃. The growth of colonies on the plate will be observed. If colonies are formed on the plate with a certain concentration and no colonies are formed on the plate with a higher concentration, the concentration is the selected concentration of drug resistance.

2. Mutagenesis treatment

(1) In order to prevent the occurrence of light-responsive mutation, the entire mutagenesis, post-mutagenesis coating and culture process are required to avoid light.

(2) Take 10 mL of single spore suspension in ϕ 90 Petri dish (with sterile magnetic stirring rod), place the dish on the magnetic stirrer in the mutagenesis chamber. The dish should be placed 30 cm directly below the UV lamp.

(3) Turn on the UV lamp, preheat for 20 mins. Turn on the magnetic stirrer and open the lid of the dish to irradiate for 15 s, 30 s, 45 s, 60 s, 75 s and 90 s respectively.

(4) Take 1. 0 mL of the control (without UV mutagenesis) and 1. 0 mL of the spore suspension treated for different times. Dilute them appropriately, take 0. 1 mL of each and spread them on the isolation medium plate. Incubate them for 2 – 3 d at 27 ℃, and calculate the cell survival rate.

3. Screening of product-tolerant mutant strains

(1) Make screening isolation plates according to the initial kanamycin tolerance of the original strain (e. g. , if the kanamycin tolerance concentration of the kanamycin-producing strain is below 1 200 μg/mL, i. e. , it does not grow on 1 200 μg/mL plates. Then 1 200 μg/mL is chosen as the kanamycin concentration of the screening isolation plates) . Take 0. 1 mL of spore suspension and 0. 1 mL of control spore suspension and apply them on the screening isolation plates. Incubate the plates at 27 ℃ for 6 – 7 d. The resistant colonies should be counted and picked onto the slant medium for storage.

(2) Make isolation medium plates containing 1 500 μg/mL, 2 500 μg/mL and 3 500 μg/mL of kanamycin, take 1 ring of the above selected strains of slant spores and scribe them in the plates. Incubate the plates at 27 ℃ for 6 – 8 d. Select 10 – 15 strains with higher tolerance to kanamycin from them and inoculate them in the slant.

(3) Inoculate the above screened resistant strains on the prepared isolation medium plates containing 10 000 μg/mL, 50 000 μg/mL and 100 000 μg/mL of kanamycin and incubate the plates at 27 ℃ for 6 – 8 d. Observe the results and select the strains with higher kanamycin toler-

ance. Inoculate the strains on the slant for preservation.

(4) The mutant strains which can tolerate higher concentration of kanamycin (10 000 μg/mL, 50 000 μg/mL, 100 000 μg/mL) will be inoculated into the fermentation medium for fermentation, and the potency of kanamycin will be measured, and the high-yielding mutant strains which can tolerate high concentration of kanamycin will be screened out from them.

【Results】

(1) Calculate the mutation rate and the amount of mutation according to the experimental results, and analyze what problems exist in the breeding process.

(2) Record the number of mutant strains obtained under different survival rates and discuss how to improve the mutation efficiency.

【Questions】

(1) What are the problems to be noted in UV mutagenesis?

(2) Why should UV mutagenesis be carried out in a stirred dish? What can other methods be used to carry out mutagenesis?

十九、细菌的亚硝基胍诱变

【实验目的】

(1) 理解并掌握亚硝基胍诱变的原理和方法。

(2) 学习筛选突变株的设计原则及方法。

【实验原理】

亚硝基胍(1 - methyl - 3 - nitro - l - nitroso - guanidin, NTG)属烷化剂,而烷化剂是突变中一类相当有效的化学诱变剂,这类诱变剂具有一个或多个活性烷基,它们易取代DNA分子中活泼的氢原子,使DNA分子上的碱基及磷酸部分被烷基化,DNA复制时导致碱基配对错误而引起突变。

NTG的作用部位一般在DNA的复制叉处。碱性时,NTG能形成重氮甲烷(CH_2N)而烷化DNA使基因发生突变。诱变机制为:重氮甲烷可导致多种碱基发生烷化,使其配对属性改变,从而诱发突变。例如,由于烷基化后的鸟嘌呤易于离子化,会由原来的酮式转化为不稳定的烯醇式,不能和胞嘧啶配对,而是与胸腺嘧啶错误配对,结果发生G:C→A:T转换而导致突变;此外,重氮甲烷还可以烷化碱基使之从DNA分子上脱落,使DNA链上的碱基空缺,在以后的复制过程中其他游离碱基就有可能错误掺入,从而造成突变;最后,重氮甲烷还可能烷化DNA中的两个鸟嘌呤,使其形成稳定共价键,从而使DNA发生交联,造成复制错误或死亡。当pH为5~5.5时,NTG能形成HNO_2,HNO_2是一种脱氨剂,可直接作用于正在复制或未复制的DNA分子,使得DNA分子上的碱基由于发生氨基的脱落而转化为其他的碱基,例如,A、C、G脱氨基后转变为H、U和X,而可与H、U和X配对的碱基分别为C、A和C,则前两种情况将导致A:T→G:C、G:C→A:T的转换突变;当pH为6.0时,NTG本身不变化,可作用于核蛋白而引起诱变效应。一般来说,NTG处理导致的突变率比其他诱变手段高,易获得高产突变株。NTG也是公认的一种超诱变剂,可以在较小的致死率对应的剂量处理后获得较大的突变幅度及突变率。

【实验仪器、材料与试剂】

1. 仪器

生化培养箱、影印模具、移液枪等。

2. 材料

菌种：枯草芽孢杆菌（*Bacillus subilis*）；牙签、锥形瓶、脱脂棉漏斗等。

3. 试剂

LB 培养基、淀粉培养基、原碘液、稀碘液、pH 6.0 磷酸缓冲液、亚硝基胍、甲酰胺。

【实验步骤】

1. 菌悬液的制备

取培养了 24 小时的枯草芽孢杆菌斜面，用 10 mL 无菌水将菌苔洗下，并倒入预先装入玻璃珠的无菌锥形瓶中，强烈振荡 10 分钟，用脱脂棉漏斗过滤，获得单细胞菌悬液。

2. 诱变处理

（1）称取 0.5 毫克 NTG 至无菌离心管中，再加入 0.05 mL 甲酰胺助溶，然后加入 0.2 mol/L pH 6.0 磷酸缓冲液 1 mL，使 NTG 完全溶解，用黑纸包裹，于 30 ℃水浴中保温平衡备用。注意，由于 NTG 见光易分解，故需要现配现用。

（2）取 4 mL 细胞悬液加入上述离心管中，充分混匀，立即置于 30 ℃水浴中振荡处理 30 分钟、40 分钟、50 分钟、60 分钟，NTG 处理终浓度为 100 μg/mL，离心收集菌体，将含有 NTG 的上清液倒入浓 NaOH 溶液中丢弃，用无菌水洗涤菌体 3 次，以终止 NTG 的诱变作用。最后向离心管中加入 5 mL 无菌水，摇匀后从中取出 0.5 mL NTG 处理后的菌悬液，用无菌水稀释至合适浓度，各取 0.2 mL 涂布淀粉培养基分离平板，30 ℃倒置培养 24 ~ 48 小时。

3. 高产淀粉酶突变株的筛选

（1）初筛。

淀粉分离平板长出菌落后，选择那些平板上有 10 ~ 30 个菌落的平板作为筛选平板，先用影印法将菌落转移至新鲜的淀粉平板或 LB 平板上，影印时注意做好标记，以便后续选择菌落，影印平板置于 30 ℃条件下倒置培养 24 小时备用。

向平板上加稀碘液数滴，摇动平板使稀碘液均匀分散在平板上，观察菌落周围出现的透明圈，分别测量透明圈直径与菌落直径并计算比值（HC 值），与对照平板（未进行诱变处理的出发菌株涂布的淀粉平板）进行比较，比值越大，说明可能产生淀粉酶的能力越强。据此初步计算致死率、突变率及突变幅度。

从影印后长好的平板上挑取那些原分离平板上 HC 值较大的菌落对应的菌株，接种于牛肉膏蛋白胨培养基中，30 ℃培养 24 小时备用。

（2）复筛。

通过初筛获得的菌株，参照标准 GB 8275 – 2009 所示的方法测定其产生淀粉酶的能力，对比原始菌株，筛选出那些产酶量高、生长速度快的菌株。

【实验结果】

（1）记录不同处理时间条件下获得的突变株的数量及致死率，讨论致死率与突变率之间的相关性。

（2）分析讨论如何提高菌株的突变率与突变幅度。

（3）将筛选出的菌株编号、保藏并写出实验报告。

【注意事项】

需要注意的是，NTG 毒性较强，操作时应小心谨慎。称量 NTG 时需佩戴手套及口罩，称量纸用完后要及时烧毁。凡是接触过 NTG 的玻璃器皿，需浸泡于 0.5 mol/L 硫代硫酸钠溶液中，置于通风处过夜，然后再用水充分洗涤。如遇到溶液外溢，用沾有硫代硫酸钠溶液的抹布擦洗，诱变处理后含 NTG 的磷酸缓冲液及稀释液，应立即倒入浓 NaOH 溶液中。若操作者皮肤接触 NTG，应立即用水冲洗。NTG 在可见光下易分解失效，故应保存于棕色瓶中。

【思考题】

（1）试述 NTG 诱变时需要注意哪些问题，为什么？

（2）在本实验中，可以利用淀粉透明圈的直径直接作为初筛指标吗，为什么？

XIX. Bacterial Mutagenesis by Nitrosoguanidine

【Objectives】

(1) Understand and master the principles and methods of nitrosoguanidine mutagenesis.

(2) Learn the design principles and methods for screening mutant strains.

【Principle】

Nitrosoguanidine (1-methyl-3-nitro-l-nitroso-guanidin, NTG) is an alkylating agent, and alkylating agents are rather effective chemical mutagens in mutagenesis. These mutagens have one or more active alkyl groups, which are more likely to replace the active hydrogen atoms in the DNA molecule, so that the bases and phosphate parts of the DNA molecule are alkylated. DNA replication leads to base pairing errors and causes mutations.

NTG usually acts at the replication fork of DNA. When in alkaline conditions, NTG can form diazomethane (CH_2N) to alkylate DNA which mutates genes. The mutagenic mechanism is: diazomethane can cause alkylation of many bases and change their pairing properties to induce mutation. For example, because the alkylated guanine is easy to ionize, it will be converted from the original ketone type to unstable enol type and cannot pair with cytosine, but with thymine wrongly. As a result, the G : C→A : F conversion will occur and lead to mutation. In addition, diazomethane can also alkylate bases to make them fall off the DNA molecule, leaving the DNA strand vacant, and other free bases may be incorrectly incorporated in the subsequent replication process, thus causing mutations. Finally, diazomethane may also alkylate two guanines in the DNA strand, making them form stable covalent bonds, thus making the DNA cross-linking occurs, resulting in replication errors or death. When the pH value is 5 – 5.5, NTG can form HNO_2, which is a deaminant that can act directly on the replicating or non-replicating DNA molecule, making the bases on the DNA molecule transform into other bases due to the occurrence of amino acid shedding. For example, A, C and G are transformed into H, U and X after deamino acid, and the bases that can pair with H, U and X are C, A and C, respectively. Then the first two cases will lead to A : T→G : C and G : C→A : T transition mutations; when the pH value is

6.0, NTG itself does not change but act on nuclear proteins to cause mutagenic effects. In general, NTG treatment leads to a higher mutation rate than other mutagenic means and is easy to obtain high-yield mutant strains. NTG is a recognized super mutagen that can obtain a large mutation amplitude and mutation rate after treatment with a dose corresponding to a smaller lethality.

【Apparatus, materials and reagents】

1. Apparatus

Biochemical incubator, photocopying mold, pipette, etc.

2. Materials

Strain: *Bacillus subtilis*; toothpicks, Erlenmeyer flasks, absorbent cotton funnel, etc.

3. Reagents

LB medium, starch medium, original iodine solution, dilute iodine solution, pH 6.0 phosphate buffer, nitrosoguanidine, formamide.

【Procedures】

1. Preparation of bacterial suspension

Take *Bacillus subtilis* slant cultured for 24 h and wash down the coating with 10 mL sterile water. Pour it into the sterile Erlenmeyer flask which is pre-loaded with glass beads, strongly vibrate for 10 mins, and filter it with absorbent cotton funnel to obtain single cell bacteria suspension.

2. Mutagenesis treatment

(1) Weigh 0.5 mg NTG into a sterile centrifuge tube, then add 0.05 mL formamide to help dissolve. Add 1 mL of 0.2 mol/L phosphate buffer (pH 6.0) to dissolve NTG completely. Wrap the tube with black paper, and keep it in a 30 ℃ water bath for equilibration. Note that NTG is easily decomposed by light, so it needs to be used immediately.

(2) Take 4 mL of cell suspension into the above centrifuge tube, mix them well, and immediately put it in the 30 ℃ water bath for 30 mins, 40 mins, 50 mins, 60 mins with shaking. The final concentration of NTG treatment is 100 μg/mL. Centrifuge the tube to collect the bacterial cells, pour the supernatant containing NTG into concentrated NaOH solution and discard it. Wash the bacterial cells with sterile water three times to terminate the mutagenesis of NTG. Finally, add 5 mL of sterile water to the centrifuge tube. Shake it well and take out 0.5 mL of NTG-treated bacterial suspension from it. Dilute it with sterile water to the appropriate concentration. Take 0.2 mL and coat it to the starch medium plate for isolation, and incubate the plate at

30 ℃ for 24 – 48 h in inverted position.

3. Screening of high yielding amylase mutant strains

(1) Primary screening.

After colonies grow in the starch isolation plate, select those plates with 10 – 30 colonies as screening plates. The colonies are transfered to a fresh starch plate or LB plate with the photocopy method. Pay attention to marking when photocopying for subsequent selection of colonies, and culture the photocopied plates upside down at 30 ℃ for 24 h.

Add several drops of dilute iodine solution to the plate. Shake the plate to evenly disperse the dilute iodine solution on the plate. Observe the transparent circle around the colony, measure the diameter of the transparent circle and the colony diameter respectively, and calculate the ratio (HC value). Compared with the control plate (starch plate coated by the original strain without mutation treatment), the higher the ratio, the stronger the production ability of amylase. Based on this, the lethal rate, mutation rate and mutation amplitude can be calculated.

The strains corresponding to the colonies with higher HC value on the original isolation plate should be selected from the plate after photocopied and inoculated in beef peptone medium, and culture at 30 ℃ for 24 hours.

(2) Re-screening.

According to the method shown in GB 8275-2009, the amylase production ability of the strains obtained through preliminary screening can be determined. Compared with the original strains, the strains with high enzyme production and fast growth rate can be selected.

【Results】

(1) Record the number and lethal rate of mutants obtained under different treatment time, and discuss the correlation between lethal rate and mutation rate.

(2) Analyze and discuss on how to improve the mutation rate and mutation amplitude.

(3) Number and store the selected strains and write the experimental report.

【Notes】

It is important to note that NTG is highly toxic and should be handled with caution. Gloves and masks should be worn when weighing NTG, and weighing paper should be burned promptly after use. Any glassware that has been exposed to NTG should be immersed in 0.5 mol/L sodium thiosulfate solution and placed in a ventilated area overnight before washing with water. If spillage of the solution is encountered, scrub with a rag soaked with sodium thiosulfate solution. The phosphate buffer and dilution solution containing NTG after mutagenesis treatment should be poured into concentrated NaOH solution immediately. If the operator's skin touches NTG, it should be rinsed with water immediately. NTG is easy to decompose and lose efficacy under visible light, so it should be kept in brown bottle.

【Questions】

(1) What problems should be paid attention to in NTG mutation and why?

(2) In this experiment, can the diameter of starch transparent ring be directly used as the initial screening index? Why?

二十、红曲的发酵及色素提取

【实验目的】

了解红曲的液体发酵方法及掌握红曲色素的提取方法。

【实验原理】

红曲色素是红曲霉在生长代谢过程中产生的天然色素，是我国食品法规允许使用的食用色素之一。红曲色素是多种色素的混合物，到目前为止，已知结构的红曲色素有10种，6种为脂溶性色素，4种为水溶性色素。脂溶性色素有红曲素、红斑素、红曲红素、红曲黄素、红斑胺和红曲红胺，水溶性色素有N－戊二酰基红斑胺、N－葡萄糖基红斑胺、N－戊二酰基红曲红胺、N－葡萄糖基红曲红胺。这些色素中，广泛应用的主要是脂溶性的红曲素、红斑素及红曲红素。红曲色素中的黄色成分约占5%，性质比较稳定，但因其含量低，所以红曲色素主要呈现红色。

红曲的生产可采用固体和液体发酵两种方法，我国传统的方法是固体发酵法，至今仍在应用。固态发酵的生产方式是将大米蒸煮、冷却后，接种红曲霉保温发酵，制成棕红色或紫红色米粒——红曲米。从本质上讲，红曲米中除了红曲色素以外，还含有未被微生物利用的米粒淀粉、淀粉酶、蛋白酶等。红曲色素也不是纯的红色素，而是红色素、橙色素与蛋白质、多肽、氨基酸等有机物的混合物，因此需要进一步纯化。

发酵得到的红曲色素为混合色素，在分光光度计上进行光谱扫描得到两个吸收峰，分别是于510 nm波长处的红色色素的吸收峰和410 nm波长处的黄色色素吸收峰，故此在测定发酵得到的红曲色素的色价时，只需测定这两处的吸收峰值即可得到总的色价。

【实验仪器、材料与试剂】

1. 仪器

恒温摇床、离心机、分光光度计、烘箱、接种钩、10 mL无菌吸管、酒精灯、10 mL具塞刻度试管、研钵等。

2. 材料

红曲霉（*M. anka*）。

3. 试剂

红曲霉斜面培养基、红曲霉种子培养基、红曲霉发酵培养基、75%乙醇溶液。

【实验步骤】

1. 发酵实验

（1）斜面培养：红曲霉转接于试管里的斜面培养基，30 ℃恒温培养 6 ~ 8 天。

（2）在液体培养基的种子培养：500 mL 三角瓶中装 100 mL 液体种子培养基。用接种钩勾取一块生长良好的红曲霉菌丝，置于液体种子培养基中，然后将种子摇瓶置于恒温摇床上培养 3 天，培养温度 30 ℃、摇床转速 250 转/分钟，偏心距 13 毫米。

（3）液体发酵培养：500 mL 三角瓶中装 100 mL 发酵培养基，按 10%接种量接种液体种子，接种后将发酵摇瓶置于恒温摇床上培养 6 天、培养温度 30 ℃、摇床转速 300 转/分钟、偏心距 13 毫米。

2. 红曲色素色价的检测

（1）发酵液色价测定。

吸取 1 mL 发酵液，置于装有 9 mL 75%乙醇溶液（pH 6.0 ~ 7.0）的具塞试管中，摇匀，充分萃取 2 小时，4 000 转/分钟离心取上清液，适当稀释后进行比色测定。

75%乙醇溶液作空白对照，用 1 厘米比色皿在分光光度计上分别测定 510 nm、410 nm 波长处的吸光值（A），吸光值乘以发酵液的稀释倍数，即为发酵液的红色色素和黄色色素的色价（U/mL），总色价为两者之和。

发酵液的色价（U/mL）$= A_{510} \times$ 稀释倍数 $+ A_{410} \times$ 稀释倍数

（2）菌丝体色价测定。

4 000 转/分钟离心或抽滤得到红曲菌丝体，80 ℃烘干菌丝体，在研钵中研磨成细粉状。准确称取 0.5 克粉末状菌丝体，用 50 mL 75%的乙醇溶液将菌丝体萃取 3 次，每次萃取 24 小时，合并萃取液，混合均匀。

取菌丝体萃取液，4 000 转/分钟离心取上清液，适当稀释后进行比色测定，测定方法同上。计算菌丝体色价（U/g）：

菌丝体色价（U/g）$= A_{510} \times$ 稀释倍数 $+ A_{410} \times$ 稀释倍数

3. 红曲霉菌丝体中红曲色素的提取

（1）抽滤红曲霉发酵液，弃去滤液，菌丝体 80 ℃烘干。

（2）在研钵中将菌丝体研磨成细粉状，也可用粉碎机粉碎成细粉状。

（3）称量菌丝体（克），加入 100 倍体积（mL）的 75%乙醇溶液。

（4）80 ℃保持 90 分钟，萃取红曲色素。

（5）过滤或抽滤除去菌丝体，然后 80℃减压浓缩得到膏状物。

（6）加入约 10 倍体积的乙酸乙酯溶解红曲膏状物，将溶解液转入分液漏斗中，

（7）静置分层。

（8）分去下层水相，60 ℃减压浓缩得到红曲粗提物。

【实验结果】

（1）描述红曲霉试管斜面、液体种子和发酵液的状态，如菌体的生长、颜色、气味等。

（2）计算发酵液和菌丝体的色价。

XX. Fermentation and Pigment Extraction of Monascus

【Objective】

Understand the liquid fermentation method of Monascus and master the extraction method of Monascus pigment.

【Principle】

Monascus pigment is a natural pigment produced by Monascus in the process of growth and metabolism. It is one of the edible pigments permitted by food regulations in China. Monascus pigment is a mixture of many kinds of pigments. So far, there are 10 kinds of Monascus pigments with known structure, 6 kinds of which are fat soluble pigments and 4 kinds of which are water soluble pigments. The fat soluble pigments are Monascus, erythema, Monascus red, Monascus flavin, erythema amine and Monascus red amine, and the water soluble pigments are N-glutaryl erythema amine, N-glucosyl erythema amine, N-glutaryl erythema amine and N – glucosyl erythema amine. Among these pigments, the most widely used pigments are fat soluble Monascus, erythema and Monascus red. The yellow component of Monascus pigment accounts for about 5%, and its properties are relatively stable, but because of its low content, Monascus pigment mainly presents red.

Solid and liquid fermentation can be used in the production of Monascus. The traditional method in China is solid fermentation, which is still in use. The production method of solid-state fermentation is that cook and cool the rice, inoculate Monascus, and keep temperature and ferment to produce brown red or purple red rice grains: Monascus rice. In essence, in addition to Monascus pigment, Monascus rice also contains rice starch, amylase, protease, etc., which are not used by microorganisms. Monascus pigment is not pure red pigment, but a mixture of red pigment, orange pigment, protein, polypeptide, amino acid and other organic compounds, so it needs further purification.

The Monascus pigment obtained by fermentation is a mixed pigment. Two absorption peaks can be observed by spectral scanning on the spectrophotometer. They are the absorption peak of red pigment at 510 nm wavelength and the absorption peak of yellow pigment at 410 nm wavelength. Therefore, when determining the color value of Monascus pigment obtained by fermentation, the total color value can be obtained just by measuring the absorption peaks of the two.

【Apparatus, materials and reagents】

1. Apparatus

Constant temperature shaker, centrifuge, spectrophotometer, oven, inoculation hook, 10 mL aseptic straw, alcohol lamp, 10 mL scaled tube with plug, mortar, etc.

2. Materials

Monascus (*M. anka*) .

3. Reagents

Aspergillus monascus slant medium, *Aspergillus monascus* seed medium, *Aspergillus monascus* fermentation medium, 75% ethanol solution.

【Procedures】

1. Fermentation experiment

(1) Inoculation in a slant medium: Monascus is transferred to the surface of the slant medium in a tube and incubated at 30 ℃ for 6 – 8 d.

(2) Seed culture in a liquid medium: 100 mL of liquid seed medium is placed in a 500 mL triangular flask. A piece of Monascus mycelium with good growth is picked up by inoculating hook and placed in the flask with liquid seed medium. Then the seed flask is placed on a constant temperature shaker for 3 days. The culture temperature is 30 ℃, the rotation speed of shaker is 250 rpm, and the eccentricity is 13 mm.

(3) Liquid fermentation culture: 100 mL of fermentation medium is added in a 500 mL triangular flask, and liquid seeds are inoculated with 10% inoculation amount. Then the flask is placed on a constant temperature shaker for 6 days. The culture temperature is 30 ℃, the rotation speed of shaker is 300 rpm, and the eccentricity is 13 mm.

2. Detection of color value of Monascus pigment

(1) Determination of color value of fermentation broth.

1 mL of fermentation broth is taken and placed in a tube with stopper containing 9 mL of 75% ethanol solution (pH 6.0 – 7.0) . Shake it well, fully extract for 2 hours. Centrifuge the supernatant at 4 000 rpm, dilute it properly and then determine the color.

75% ethanol solution is used as blank control. The absorbance values (A) at 510 nm wavelength and 410 nm wavelength are measured by spectrophotometer with 1 cm cuvette. The absorbance value multiplied by the dilution factor of fermentation broth is the color value (U/mL) of red pigment and yellow pigment in fermentation broth, and the total color value is the sum of the two.

The color value of fermentation broth (U/mL) = A_{510} × dilution factor + A_{410} × dilution factor.

(2) Determination of mycelium color value.

After centrifugation at 4 000 rpm or filtration, Monascus mycelium is dried at 80 ℃ and ground into fine powder in mortar. Weigh accurately 0. 5 g of powdered mycelium; extract it three times with 50 mL of 75% ethanol solution for 24 h each time. Collect all the extract and mix it evenly.

The extract of mycelium is centrifuged at 4 000 rpm, and the supernatant is diluted properly for colorimetric determination. The determination method is the same as above. The color value (U/g) of mycelium is calculated as followed:

Mycelium color value (U/g) = A_{510} × dilution factor + A_{410} × dilution factor

3. Extraction of Monascus pigment from Monascus mycelium

(1) The fermentation broth of *Monascus filterum* is filtrated, the filtrate is discarded, and the mycelium is dried at 80 ℃.

(2) Grind mycelium into fine powder in a mortar or crush it into fine powder by a grinder.

(3) Weigh mycelium (g) and add 100 times volume (mL) of 75% ethanol solution.

(4) Extract Monascus pigment at 80 ℃ for 90 mins.

(5) Remove mycelium by filtration or suction, and then vacuum concentrate it to obtain paste at 80 ℃.

(6) Add about 10 times volume of ethyl acetate to dissolve the Monascus paste, and transfer the solution into the separation funnel.

(7) Stand for stratification.

(8) Separate the lower aqueous phase and concentrate at 60 ℃ under reduced pressure to obtain the crude extract of Monascus.

【Results】

(1) Describe the culture condition of Monascus in tube slopes, liquid seeds, and fermentation broth, such as growth, color, gas taste, etc.

(2) Calculate color values of fermentation broth and mycelium.

二十一、谷氨酸的发酵及产物提取

【实验目的】

（1）熟悉谷氨酸发酵的工艺流程、了解发酵过程参数变化及控制。

（2）掌握等电点法提取谷氨酸的方法。

【实验原理】

谷氨酸是一种两性电解质，等电点 pI 3.22，主要作为味精（即谷氨酸钠）的生产原料。微生物将三羧酸循环中产生的 α－酮戊二酸在谷氨酸脱氢酶作用下还原氨化成谷氨酸。谷氨酸产生菌生物合成谷氨酸途径包括糖酵解途径（EMP）、磷酸己糖旁路途径（HMP）、三羧酸循环（TCA 循环）、乙醛酸循环、伍德－沃克曼反应（CO_2的固定反应）等，其合成的理想途径为：葡萄糖先经 EMP 生成丙酮酸，丙酮酸经丙酮酸脱氢酶复合酶系催化作用生成乙酰 CoA，乙酰 CoA 进入三羧酸循环生成 α－酮戊二酸，α－酮戊二酸再经氨基化作用生成谷氨酸。尽管谷氨酸产生菌的理想代谢途径比较简单，但是复杂的代谢调控系统需在生产过程中严格调控。

发酵条件中对微生物发酵生产谷氨酸的影响因素如下：

（1）温度：谷氨酸发酵前期（0～12 小时）为菌体生长繁殖阶段，此阶段菌体利用培养基中营养物合成核酸、蛋白质等，谷氨酸产生菌最适生长温度为 30 ℃～32 ℃，在发酵中后期是谷氨酸大量积累的阶段，催化谷氨酸合成的谷氨酸脱氢酶的最适温度为 32 ℃～36 ℃。

（2）pH：谷氨酸脱氢酶是谷氨酸合成的主要酶系，合成谷氨酸脱氢酶的最适 pH 为 7.0～7.2。当发酵液 pH 偏酸时（pH 5.0～5.8），谷氨酸脱氢酶活性受到抑制，代谢生成谷氨酰胺；发酵后期由于合成谷氨酸消耗大量的 NH_4^+，pH 下降，此时需要进行 pH 调节以保证发酵的正常进行，发酵中后期通过添加尿素、氨水等方法调节 pH 至 7.0～7.6。

（3）生物素：产谷氨酸的微生物都为生物素营养缺陷型菌株，生物素供应充足则发酵菌体大量繁殖，少产生谷氨酸，因此在细胞生长阶段给予充足生物素。谷氨酸产生阶段生物素需控制在亚适量 5～10 μg/L，才能使谷氨酸大量积累。

（4）通气：细胞生长阶段通气速率需适中，否则高通气会导致细胞大量生长，造成底物代谢大部分流向细胞生长而非代谢产物合成；但谷氨酸合成阶段必须大量通气，否则会生成乳酸或琥珀酸。

（5）NH_4^+浓度：NH_4^+浓度可影响发酵液的 pH，合适的 NH_4^+促进 α－酮戊二酸转化成谷氨酸，NH_4^+过量则会导致生成的谷氨酸进一步氨化成谷氨酰胺；NH_4^+不足则无法使 α－酮戊二酸氨化转化成谷氨酸。

（6）碳源浓度：碳源浓度在一定范围内，谷氨酸随着碳源浓度的升高而增加，当碳源浓度过高时，渗透压过大，对菌体生长不利，从而使谷氨酸对碳源的转化率降低。因此发酵液中的碳源的浓度控制在 10% ~13%.

（7）碳氮比：谷氨酸合成中需要大量的氨，约 85% 的氮源被用于谷氨酸的合成，谷氨酸合成过程中的碳氮比一般为 100 ：（15 ~21），碳氮比增加导致菌体生长而不合成谷氨酸。

谷氨酸为两性电解质，在等电点处溶解度最小，因此可采用等电点方法提取谷氨酸，将发酵液 pH 调节至谷氨酸等电点（pI 3.22），使谷氨酸呈现过饱和状态而析出。

【实验仪器、材料与试剂】

1. 仪器

7.5 L 自动控制发酵罐、高压蒸汽灭菌锅、SBA－40 生物传感器、恒温培养箱、振荡培养箱、天平、超净工作台、分光光度计、高速离心机、显微镜、旋转蒸发器、恒温水浴锅、三角瓶、烧杯、量筒、玻璃棒、pH 计、容量瓶、移液器及枪头、离心管等。

2. 材料

北京棒状杆菌（*Corynebacterium pekinense*）：生物素缺陷型。

3. 试剂

北京棒状杆菌斜面培养基、北京棒状杆菌一级种子培养基、北京棒状杆菌二级种子培养基、北京棒状杆菌发酵培养基、硅藻土。

【实验步骤】

1. 菌种活化

将北京棒杆菌从－80 ℃冰箱保存的甘油管中挑取一环培养物，划线于斜面培养基上，32 ℃培养至菌落直径 2 毫米备用。

2. 种子制备

（1）一级种子：从斜面培养基上挑取一环菌种接种于 100 mL 一级种子培养基中，于 32 ℃ 200 转/分钟振荡培养 12 小时。一级种子质量要求为：无杂菌、无噬菌体、pH 6.4 ±0.1。

（2）二级种子：取 10 mL 一级种子接种到二级种子培养基中，于 33 ~34 ℃ 200 转/分钟振荡培养7~8小时。二级种子质量要求：无杂菌、无噬菌体、pH 7.0±0.2、残糖消耗1%左右、镜检生长旺盛、排列整齐。

3. 发酵培养基灭菌及接种

配制好5 L发酵培养基装入到7.5 L发酵罐内，配置400 mL浓度为40%的尿素及30 mL消泡剂。121 ℃灭菌30分钟，灭菌结束后通入冷却水使培养基降温30 ℃~32 ℃，将二级种子按照接种量2%接种到发酵罐中。

4. 发酵过程控制

(1) 温度控制：0~6小时温度控制为33 ℃，7~25小时每隔6小时升温1 ℃；26~34小时，温度控制为37~38 ℃.

(2) pH控制：采用流加氨水控制pH。发酵前期pH 7.0，后期提到pH 7.2~7.3以保证合成谷氨酸所需的氮源；发酵后期pH稍微降低至约7.0；在发酵结束前6小时停止流加氨水，放罐时pH约6.5。

(3) OD净增值控制：总OD净增值控制在0.75~0.8。

(4) 通气量控制：发酵开始通气量为1：0.08；OD净增值为0.25时，通气量为1：0.15；OD净增值为0.50时，通气量为1：0.18；OD净增值>0.65时，通气量为1：0.24，保持约10小时。

(5) 泡沫控制：配制20%~30%浓度的消泡剂，经灭菌、冷却后备用。在发酵产生一定泡沫时，流加一定量的消泡剂，每次流加10 mg/L。

(6) 放罐指标：当发酵周期达到32~35小时、残糖<10 g/L、谷氨酸含量>65 g/L时可以放罐。使用生物传感器测定谷氨酸和葡萄糖含量。

5. 谷氨酸的等电点法提取

(1) 将100 mL发酵液加盐酸调节pH至4.0~4.5（起晶中和点），投入谷氨酸晶种0.1克，育晶搅拌3~4小时。

(2) 继续加盐酸调节pH至3.2，pH调节速率为0.1 pH/小时，搅拌20小时后停止搅拌，沉降3~4小时，5 000转/分钟离心分离沉淀即得到谷氨酸粗品。

(3) 母液用盐酸调节pH至2，加硅藻土0.5克，加热至70 ℃放置沉降5~6小时，过滤机过滤. 滤液温度降至50 ℃时缓慢加入碱液中和，待晶核形成（30~40分钟）再加碱中和至pH 3.22，搅拌24小时，沉降分离谷氨酸（二次谷氨酸）。

(4) 混合步骤（2）和（3）得到的谷氨酸湿物料，经旋转、蒸发、干燥后得到谷氨酸粗品。

【实验结果】

发酵过程控制参数和结果记录于表21-1。

表 21－1　发酵过程控制参数及结果记录表

时间/h	温度/℃	pH	OD 值	通气量/(L·min^{-1})	OD_{600}	葡萄糖含量	谷氨酸含量
0							
4							
8							
12							
16							
20							

XXI. Fermentation and Extraction of Glutamic Acid

【Objectives】

(1) Be familiar with glutamic acid fermentation process and understand the changes and control of parameters in the fermentation process.

(2) Master the method of extracting glutamic acid by isoelectric point method.

【Principle】

Glutamate is an amphoteric electrolyte with isoelectric point (pI) of 3. 22, which is mainly used as the raw material of monosodium glutamate (MSG) production. α-ketoglutarate acid produced in the tricarboxylic acid cycle is reduced to glutamate by microorganism under the action of glutamate dehydrogenase. Glutamic acid biosynthesis pathway of glutamic acid producing bacteria includes glycolysis pathway (EMP), hexose phosphate bypass pathway (HMP), tricarboxylic acid cycle (TCA cycle), glyoxylic acid cycle, Wood-Volkman reaction (CO_2 fixation reaction), etc. The ideal pathway of the synthesis is: glucose first generates pyruvate through EMP. Pyruvate generates acetyl CoA through the catalysis of pyruvate dehydrogenase complex enzyme system. Then acetyl CoA enters the tricarboxylic acid cycle to generate α-ketoglutarate, and α-ketoglutarate generates glutamate through amination. Although the ideal metabolic pathway of glutamic acid producing bacteria is relatively simple, the complex metabolic regulation system needs to be strictly regulated in the production process.

The factors influencing the production of glutamic acid by microbial fermentation are as follows.

(1) Temperature: The prophase of glutamic acid fermentation (0 – 12 h) is the stage of cell growth and reproduction. In this stage, the cell uses nutrients in the medium to synthesize nucleic acid, protein and so on. The optimal growth temperature of glutamic acid producing bacteria is 30 ℃ – 32 ℃; the optimum temperature of glutamate dehydrogenase synthesis is 32 ℃ – 36 ℃ in the middle and late stages of fermentation, when glutamate accumulates.

(2) pH: Glutamate dehydrogenase is the main enzyme of glutamate synthesis system, and

the optimal pH of glutamate dehydrogenase synthesis is 7.0 – 7.2. When the pH of the fermentation broth is slightly acidic (pH 5.0 – 5.8), the activity of glutamate dehydrogenase is inhibited and glutamine is metabolized; in the late stage of fermentation, due to the consumption of a large amount of NH_4^+ in the synthesis of glutamate, the pH decreases, so it is necessary to adjust the pH to ensure the normal fermentation. In the middle and late stages of fermentation, the pH is adjusted to 7.0 – 7.6 by adding urea, ammonia and other methods.

(3) Biotin: Glutamate-producing microorganisms are all biotin deficient strains. If the supply of biotin is sufficient, the fermentation cells will multiply in large quantities and produce less glutamic acid. Therefore, sufficient biotin is given in the cell growth stage. At the stage of glutamate production, biotin should be controlled at a sub appropriate amount of 5 – 10 μg/L to accumulate glutamate.

(4) Ventilation: During the cell growth stage, the ventilation rate should be moderate, otherwise high ventilation will lead to a large number of cell growth, causing most of the substrate metabolism supplied to the cell growth rather than the synthesis of metabolites; however, during the glutamate synthesis stage, enough ventilation is required, otherwise lactic acid or succinic acid will be generated.

(5) NH_4^+ concentration: The concentration of NH_4^+ can affect the pH of fermentation broth. Appropriate NH_4^+ can promote the conversion of α-ketoglutarate to glutamic acid. Excessive NH_4^+ will lead to the further ammoniation of glutamic acid to glutamine; insufficient NH_4^+ can not make the ammoniation of α-ketoglutarate to glutamic acid.

(6) Carbon source concentration: When the carbon source concentration is within a certain range, glutamic acid increases with the increase of carbon source concentration. When the carbon source concentration is too high, the osmotic pressure is too high, which is unfavorable to the growth of bacteria, thus reducing the conversion rate of glutamic acid to carbon source. Therefore, the concentration of carbon source in fermentation broth should be controlled at 10% – 13%。

(7) C/N ratio: A large amount of ammonia is needed in the synthesis of glutamic acid; about 85% of the nitrogen source is used for the synthesis of glutamic acid. The C/N ratio in the process of glutamic acid synthesis is generally 100 : (15 – 21), and the increase of C/N ratio leads to the growth of bacteria without glutamic acid synthesis.

Glutamic acid is an amphoteric electrolyte with minimum solubility at the isoelectric point. Therefore, the isoelectric point method can be used to extract glutamic acid, and the pH of fermentation broth can be adjusted to the isoelectric point of glutamic acid (pI 3.22) to make glutamic acid supersaturate and precipitate.

【Apparatus, materials and reagents】

1. Apparatus

7.5 L automatic control fermenter, autoclave, SBA-40 biosensor, constant temperature incubator, oscillating incubator, balance, ultra-clean worktable, spectrophotometer, high-speed centrifuge, microscope, rotary evaporator, constant temperature water bath, Erlenmeyer flasks, beakers, measuring cylinders, glass rods, pH meters, volumetric flasks, pipettes and spear heads, centrifuge tubes, etc.

2. Materials

Corynebacterium pekinense: biotin deficient type.

3. Reagents

Slant medium for *Corynebacterium pekinense*, primary seed medium of *Corynebacterium pekinense*, secondary seed medium of *Corynebacterium pekinense*, fermentation medium of *Corynebacterium pekinense* and diatomite.

【Procedures】

1. Activation of strains

A ring of culture is selected from the containing glycerol tube stored in refrigerator at −80 ℃, and then scribed on the slant medium. The culture should be kept at 32 ℃ until the colony diameter reached 2 mm for use.

2. Seed preparation

(1) Primary seed: A ring of bacteria is picked up from the slant culture medium and inoculated in 100 mL primary seed medium. The bacteria should be shaken at 32 ℃ 200 rpm for 12 h. The primary seed quality requirements are: no bacteria, no phage, pH 6.4 ±0.1.

(2) Secondary seed: 10 mL of primary seed culture are inoculated into the medium of secondary seed and cultured at 33 ℃ −34 ℃ and 200 rpm for 7 −8 h with shaking. Secondary seed quality requirements are: no bacteria, no phage, pH 7.0 ±0.2, about 1% of residual sugar consumption, vigorous growth under microscopic examination, orderly arrangement.

3. Sterilization and inoculation of fermentation medium

5 L fermentation medium is prepared and put into a 7.5 L fermenter. Prepare 400 mL of 40% urea and 30mL defoamer. After sterilization at 121 ℃ for 30 mins, cool the medium to 30 ℃ −32 ℃ with cooling water, and inoculate the secondary seeds into the fermenter with 2% inoculation amount.

4. Fermentation process control

(1) Temperature control: The temperature is controlled at 33 ℃ in 0 – 6 h and increased by 1 ℃ every 6 h in 7 – 25 h. The temperature is 37 ℃ – 38 ℃ in 26 – 34 h.

(2) pH control: The pH is controlled by adding ammonia water. In the early stage of fermentation, the pH is 7.0; and in the later stage, the pH is 7.2 – 7.3 to ensure the nitrogen source needed for the synthesis of glutamic acid. In the lastest stage of fermentation, the pH is slightly reduced to about 7.0; at the end of fermentation, the flow of ammonia water is stopped 6 hours before the end of fermentation, and the pH was about 6.5 when putting into the fermenter.

(3) OD net increment control: The total OD net increment is controlled between 0.75 – 0.8.

(4) Ventilation control: At the beginning of fermentation, the ventilation rate is 1 : 0.08; when the OD net increment is 0.25, the ventilation rate was 1 : 0.15; when the OD net increment is 0.50, the ventilation rate is 1 : 0.18; when the OD net increment is more than 0.65, the ventilation rate is 1 : 0.24 for about 10 h.

(5) Foam control: The defoamer with concentration of 20% – 30% is firstly prepared and used after sterilization and cooling. When a certain amount of foam is produced, defoamer is added. 10 mg/L of defoamer is added every time.

(6) Indexes of opening fermenter: When the fermentation period is 32 – 35 h, residual sugar is less than 10 g/L and glutamic acid content is more than 65 g/L, the fermenter can be stopped and opened. The contents of glutamic acid and glucose can be determined by biosensor.

5. Extraction of glutamic acid by isoelectric point method

(1) The pH of 100 mL fermentation liquid is adjusted to 4.0 – 4.5 (crystallization at neutralization point) with hydrochloric acid. 0.1 g glutamate crystal seed is put into the fermentation liquid. Stir for 3 – 4 h for crystallization.

(2) Continue to add hydrochloric acid to adjust pH to 3.2. Adjust pH at a rate of 0.1 pH/h. Then stop stirring after 20 h of stirring. Precipitate for 3 – 4 h and centrifuge at 5 000 rpm to obtain a crude product of glutamic acid.

(3) The pH of mother liquor is adjusted to 2 with hydrochloric acid, followed by adding 0.5 g diatomite. Heat the mother liquor to 70 ℃ and precipitate for 5 – 6 h. Filter and collect the filtrate. When the temperature of filtrate drop to 50 ℃, alkaline solution is added slowly for neutralization. After crystal nuclear forms (30 – 40 mins), alkaline solution is further added for neutralization to pH 3.22. Stir for 24 h, and glutamic acid (secondary glutamic acid) is separated by sedimentation.

(4) The wet products of glutamic acid obtained in steps (2) and (3) are mixed, and the

crude glutamic acid is obtained after rotating, evaporating and drying.

【Result】

Record the control parameters and results of fermentation process in Table XXI – 1.

Table XXI – 1 The control parameters and results of fermentation process

Time/h	Temperature/℃	pH	The OD value	Ventilation/ (L · min^{-1})	OD_{600}	Glucose content	Glutamate content
0							
4							
8							
12							
16							
20							

二十二、培养材料的消毒与接种

【实验目的】

初步掌握材料消毒、接种的方法和技术。

【实验原理】

初次接种的材料，表面都带有各种微生物，必须在接种前对其进行消毒。消毒的基本原则是既要把材料表面附着的微生物杀死，又不伤害材料内部的组织、细胞。因此，消毒所采用药剂的种类、浓度、处理时间的长短，均应根据材料的种类、组织的老嫩、茸毛的有无及材料对药剂的敏感性来定。

【实验仪器、材料与试剂】

1. 仪器与设备

超净工作台、镊子、手术刀、剪刀、酒精灯、棉球、火柴、烧杯、废液缸、无菌滤纸、培养皿（无菌）、解剖镜。

2. 材料

作物、果树、蔬菜、花卉等植物的茎尖、带腋芽的嫩茎及嫩叶等。

3. 试剂

培养基（根据不同植物器官和组织材料选择基本培养基）、0.1%升汞（或2%次氯酸钠）、70%酒精、无菌水等。

【实验步骤】

1. 接种前的准备

（1）培养基的准备。

按培养材料的要求，选取合适的各种培养基，待用。

（2）超净工作台的准备。

在超净工作台上摆好酒精灯、一瓶70%酒精棉球、无菌培养皿、无菌滤纸、镊子、手术刀、火柴、0.1%升汞（或2%次氯酸钠）、一瓶70%酒精、无菌水、培养基等。

无菌室（包括缓冲间）用紫外灯灭菌 15 ~ 20 分钟。超净工作台开机过滤空气并开紫外灯 15 ~ 20 分钟，然后关掉紫外灯，用 70% 酒精喷雾降尘。

（3）培养材料的准备。

从田间或温室选取生长旺盛无病虫害的顶芽、幼嫩叶片和嫩茎段（去掉老叶，剪成合适的长度），放入烧杯中；自来水反复漂洗干净（可适当加几滴洗涤剂漂洗），备用。

2. 培养材料的消毒

（1）操作前先用肥皂洗手，再用 70% 酒精棉球将手（特别是指尖）、手臂、工作台面，放在台面上的所有操作用具、器皿等都擦一遍，以清除尘粒。点燃酒精灯，把操作用的器械（镊子、手术刀等）放在酒精灯上灼烧灭菌，之后放在支架上冷却待用。

（2）将材料先倒入 70% 酒精浸泡，摇动几下，10 ~ 20 秒后立即倒出酒精，加无菌水漂洗，然后倒入 0. 1% 升汞（或 2 % 次氯酸钠）进行表面消毒，时间的长短视材料而定，一般是 5 ~ 12 分钟或更长（常用消毒剂及消毒时间见表 22 – 1）。药剂浸泡过程中应不断摇动，然后用无菌水冲洗 3 ~ 5 遍，备用。

（3）对于果树或木本花卉未萌动或刚刚萌动的芽，一般可进行两次灭菌。即在上述过程完成后，进行鳞片及幼叶剥离，然后再进行第二次消毒，此次消毒时间宜短，用无菌水冲洗 3 ~ 5 遍后剥取茎尖接种。

表 22 – 1　常用消毒剂和消毒时间

消毒剂	使用质量	去除难易	消毒时间/分钟	消毒效果	是否毒害植物
次氯酸钙	9 ~ 10	易	5 ~ 30	很好	低毒
次氯酸钠	2	易	5 ~ 30	很好	无
过氧化氢	10 ~ 12	最易	5 ~ 15	好	无
硝酸银	1	较难	5 ~ 30	好	低毒
氯化汞	0. 1 ~ 1	较难	2 ~ 10	最好	剧毒
酒精	70 ~ 75	易	0. 2 ~ 2	好	有
抗生素	4 ~ 50（mg/L）	中	30 ~ 60	很好	低毒

3. 无菌接种

经消毒后的材料，必须在无菌条件下接种到培养基中，进行无菌培养。具体操作如下：

（1）培养基按成分不同分类摆放在工作台边上。注意不要挡住无菌风。

（2）轻轻打开瓶盖，将瓶口迅速在酒精灯火焰上方转动一圈灼烧灭菌，然后放在工作台上。

（3）用镊子取出待接种材料置于垫有无菌滤纸的培养皿内，进行材料的剥离和切割。

①茎段：一般切成带一个芽（芽最好位于中央）的茎段，两端切掉被消毒剂杀伤的切口，接种到所选培养基上，腋芽要露在培养基之上，且不要倒置。每瓶接种 3~4 个茎段。再将瓶口和瓶盖在火焰上方迅速旋转灼烧灭菌，盖紧瓶盖或封口膜，注明名称、接种日期。

②幼叶：用刀切去叶缘，叶脉，将叶片切成约 5 毫米×5 毫米的小块，接种到所选培养基上，每瓶接种 10~12 块。瓶口和瓶盖灼烧灭菌后盖紧，注明名称、接种日期。

③茎尖：在解剖镜下，刺取直径为 0.2~0.5 毫米带有 1~2 个叶原基的茎尖分生组织，置所选培养基表面进行培养，每瓶放置 3~4 个。封口，注明名称、接种日期。

注意：为了防止交叉感染，工具要用一次即灼烧灭菌一次，滤纸和培养皿也要及时更换。操作人员手指也要经常用 70% 酒精棉球擦拭。在接种过程中严禁讲话、咳嗽和走动。

（4）无菌操作完成后，关闭工作台电源，把材料转入光照培养室进行培养，把废液倒入回收桶处理，清理干净工作台，关上挡风玻璃。

4. 培养条件与培养室消毒

（1）培养条件：在植物组织培养中，培养室温度通常控制 25±2 ℃，每日光照 12~16 小时，光照强度 1 000~5 000 lx，相对湿度 70%~80%。

（2）培养室消毒：可用紫外灯杀菌，每次杀菌 30 分钟。紫外灯使用久了杀菌效果会降低，要相对延长杀菌时间。到了使用寿命要及时更换，否则影响杀菌效果。

也可用甲醛氧化浓氨法对培养室消毒。即一次投以足量的甲醛及氧化剂（高锰酸钾），通过氧化作用，迅速放出高浓度的甲醛气体杀灭微生物。在熏蒸时间达 4 小时后，利用浓氨气体与氧化反应剩余的甲醛分子发生化学反应，生成无毒无刺激性的甲胺，以减少有毒气体的伤害。此法操作简便，杀菌效果极好，但在使用时，应准确掌握用量。甲醛用量为 320 mL/m^3，一次熏蒸时间不少于 4 小时，浓氨水用量与甲醛的比例为 1∶4，加入浓氨水后熏蒸 30 分钟。

【注意事项】

升汞（$HgCl_2$）为剧毒药品，一则使用时注意别溅到皮肤上，二则使用后要倒入回收桶内进行处理，不能直接倒入下水道。升汞处理方法如下：

$HgCl_2 + Na_2S \rightarrow HgS$ 絮状沉淀（至絮状沉淀不再产生为止）。

$Na_2S + FeSO_4 \rightarrow FeS + Na_2SO_4$（中和过多的 NaS）。

【思考与记录】

填写下列调查表。

接种污染调查表

调查日期　　年　　月　　日

材料名称	接种日期	接种数量/瓶	污染数量/瓶	污染率/%	污染菌种类型

XXII. Disinfection and Inoculation of Living Material

【Objective】

Learn the methods and techniques of material disinfection and inoculation.

【Principle】

The initial inoculation materials, with various microbes on the surface, must be disinfected before inoculation. The basic principle of disinfection is to kill the microbes attached to the surface of the material without harming the tissues and cells inside the material. Therefore, the type, concentration and treatment time of the medicament used in disinfection should be determined according to the type of material, the tenderness of the tissue, the presence or absence of fuzz and the sensitivity of the material to the medicament.

【Apparatus, materials and reagents】

1. Apparatus and equipment

Ultra-clean worktable, tweezers, scalpel, scissors, alcohol lamp, cotton ball, match, beaker, waste liquid cylinder, sterile filter paper, Petri dishes (sterile) and anatomical mirror.

2. Materials

Stem tips of plants such as crops, fruit trees, vegetables and flowers, or tender stems and leaves with axillary buds.

3. Reagents

Medium (select basic medium according to different plant organ and tissue materials), 0.1% mercury dichloride (or 2% sodium hypochlorite), 70% alcohol, aseptic water, etc.

【Procedures】

1. Preparation before inoculation

(1) Preparation of medium.

According to the requirements of culture materials, select appropriate medium for use.

(2) Preparation of the ultra-clean worktable.

Put an alcohol lamp, a bottle of cotton balls with 70% alcohol solution, sterile Petri dishes, sterile filter paper, tweezers, a scalpel, a box of match, 0.1% mercuric chloride (or 2% sodium hypochlorite), a bottle of 70% alcohol, sterile water, medium and so on on the ultra-clean table.

Sterilize the sterile room (including buffer room) by UV lamp for 15 – 20 mins. Filter air and turn on UV lamp for 15 – 20 mins. Then turn off UV lamp and spray 70% ethanol to precipitate dust.

(3) Preparation of culture materials.

Select the terminal buds, young leaves and tender stems (remove the old leaves and cut into the appropriate length) from the field or greenhouse and put them in a beaker; rinse by tap water repeatedly (add a few drops of detergent to rinse properly) and set aside.

2. Disinfection of culture materials

(1) Wash hands with soap before operation, then use cotton ball with 70% alcohol to wipe hands (especially fingertips), arms, worktable and all operating utensils on the table to remove dust particles. Ignite the alcohol lamp, put the instrument (tweezers, scalpel, etc.) on the alcohol lamp to burn and sterilize, then put it on the support to cool and wait for use.

(2) Soak the material with 70% alcohol solution, shake a few times for 10 – 20 s. Pour out the alcohol immediately, rinse the material with sterile water. After that, spray 0.1% mercury dichloride (or 2% sodium hypochlorite) onto the material for surface disinfection. Generally 5 – 12 mins or longer (common disinfectants and their disinfection time listed in Table XXII – 1). Shake it continuously during soaking, then rinse with aseptic water 3 – 5 times for use.

(3) The buds of fruit trees or woody flowers that have not sprouted or just sprouted can be sterilized twice. That is, after the completion of the above process, the scales and young leaves should be peeled, and then the second disinfection should be carried out. The disinfection time should be short, and the stem tips should be peeled and inoculated after washing with sterile water for 3 – 5 times.

Table XXII – 1 Common disinfectants and disinfection time

Disinfectant	Quality of use	Difficulty of removal	Disinfection time/min	Disinfection effect	Is it toxic to plants
Calcium hypochlorite	9 – 10	Easy	5 – 30	Great	Low toxicity
Sodium hypochlorite	2	Easy	5 – 30	Great	—
Hydrogen peroxide	10 – 12	Most easily accessible	5 – 15	Good	—
Silver nitrate	1	More difficult	5 – 30	Good	Low toxicity
Mercury	0.1 – 1	More difficult	2 – 10	Best	Highly toxicity
Alcohol	70 – 75	Easy	0.2 – 2	Good	Normal toxicity
Antibiotics	4 – 50 (mg/L)	Medium	30 – 60	Great	Low toxicity

3. Aseptic inoculation

The sterilized materials must be inoculated into the medium under aseptic conditions for aseptic culture. The operation is as follows:

(1) Media are placed on the edge of the worktable according to the different composition. Be careful not to block the aseptic wind.

(2) Gently open the cap, turn the bottle mouth quickly over the flame of the alcohol lamp and sterilize it, then put it on the worktable.

(3) The material for inoculation is removed with tweezers and placed on a Petri dish with aseptic filter paper for material stripping and cutting.

①Stem segment: a stem segment with a bud (preferably in the center) is usually taken. Cut off the incision damaged by disinfectant at both ends. Inoculate the material on the selected medium. Axillary buds should be exposed on the medium, and not inverted. Inoculate 3 – 4 stem segments in each bottle. Then the bottle mouth and cap are quickly rotated over the flame for sterilization, and the cap or sealing film is tightly closed. The name and inoculation date is labeled.

②Young leaves: cut the leaf margin and veins with a knife. Cut the leaves into small pieces of about 5 mm × 5 mm, inoculate them on the selected medium. 10 – 12 pieces are inoculated in each bottle. The mouth and cap of the bottle are burned for sterilization and then covered tightly. Lable the name and inoculation date on it.

③Stem tip: the apical meristem with 1 – 2 leaf primordia and 0.2 – 0.5 mm in diameter are obtained and cultured on the surface of the selected medium, 3 – 4 in each bottle. Seal and label the name and inoculation date.

Note: in order to prevent cross-infection, tools should be sterilized once after using, and filter paper and Petri dish should be replaced in time. Operators should also often use cotton balls with 70% alcohol to wipe fingers. It is forbidden to speak, cough and walk during inoculation.

(4) After aseptic operation, turn off the power supply of the worktable. Transfer the material to the light culture room for cultivation. Pour the waste liquid into the recycling bucket for treatment, clean the workbench and close the windshield.

4. Culture conditions and disinfection of culture room

(1) Culture conditions: in plant tissue culture, the temperature of culture room is usually controlled at 25 ℃ ± 2 ℃, the time of illumination is 12 – 16 h per day, the light intensity is 1 000 – 5 000 lx, and the relative humidity is 70% – 80%.

(2) Disinfection of culture room: ultraviolet lamp is used to sterilize the culture room for 30 mins. UV lamp used for a long time will reduce the sterilization effect. If so, it needs to relatively extend the sterilization time. When the service life of the ultraviolet lamp is up, it should be replaced in time, otherwise the sterilization effect will be affected.

The culture room can also be disinfected by the method of formaldehyde oxidation and concentrated ammonia. That is to say, a sufficient amount of formaldehyde and oxidant (potassium permanganate) are applied at one time, and high concentration of formaldehyde gas is rapidly released through oxidation to kill microorganisms. After fumigation for 4 hours, the concentrated ammonia gas reacts with the residual formaldehyde molecules by the oxidation reaction to produce nontoxic and non irritating methylamine, so as to reduce the damage of toxic gas. This method is easy to operate and has excellent germicidal efficacy, but the dosage should be accurately controlled. The dosage of formaldehyde is 320 mL/m^3, and the fumigation time should be not less than 4 h. The ratio of concentrated ammonia to formaldehyde is 1 : 4; and the fumigation time should be 30 mins after adding concentrated ammonia.

【Notes】

Mercury dichloride ($HgCl_2$) is a highly toxic drug. Don't splash on the skin during operation; pour it into the recycling bucket for treatment, do not pour directly into the sewer. The mercury dichloride treatment is as follows:

$HgCl_2 + Na_2S \rightarrow HgS$ Flocculent precipitation (until flocculent precipitation is no longer produced).

$Na_2S + FeSO_4 \rightarrow FeS + Na_2SO_4$ (neutralizing excessive Na_2S).

【Thinking and recording】

Fill in the following questionnaire.

Questionnaire on inoculation contamination

Date of investigation: ____________

Material name	Date of inoculation	Number of inoculations/bottle	Quantity of contamination/bottle	Pollution rate/%	Type of contaminated strain

二十三、试管苗的驯化移栽

【实验目的】

（1）掌握试管苗的生长环境和特性。

（2）掌握试管苗驯化炼苗移栽的方法、炼苗基质的配备和炼苗期间的管理技术。

【实验原理】

试管苗的获得和生根只是组培工作的一部分。驯化炼苗是试管苗与大田生产苗的重要衔接，试管苗从瓶内到瓶外，由恒温、高湿、弱光、无菌的生活环境转换到变温、变湿、有菌的生活环境，变化十分剧烈，而且试管苗植株幼嫩，表皮角质层薄，抵抗力弱。因此，在移栽大田前先要驯化炼苗，炼苗初期空气应保持一定温度、湿度，炼苗基质要肥沃、通气、湿润。炼苗成活率直接影响到前期工作的成本和后期大田的定植。

【实验仪器、材料与试剂】

1. 仪器与设备

炼苗盘、温室、塑料大棚。

2. 材料

试管苗、草炭土、珍珠岩、蛭石、沙壤土、营养体。

【实验步骤】

1. 炼苗基质的配备

不同的植物需要不同的炼苗基质。要求炼苗基质肥沃通透性好，利于透水和根系的生长，这是提高炼苗成活率的基本条件。调配基质时要用稀释 500～800 倍的多菌灵或甲基托布津等杀菌剂喷雾消毒（实验室少量基质可用高压锅灭菌），用雾状水拌匀基质，基质湿度以 60% 左右为宜，即手握成团，落地散开，不要湿度过大，以免小苗出现烂根、烂茎等现象。基质拌匀装钵后备用，存放时间不要超过半天。

一般植物使用草炭土、珍珠岩、蛭石、沙壤土，经消毒灭菌后，按一定的比例混合，不同的季节稍有变动。一般植物常用的基质配比：

①草炭土：珍珠岩：蛭石 =3∶1∶1。

②草炭土：珍珠岩 =2∶1（夏季炼苗用，因蛭石容易保水提温）。

③草炭土：沙壤土 =1∶1。

兰花类可选用海苔草作为炼苗基质。

2. 试管苗驯化移栽

当植株在生根培养基中长成 4 ~ 5 厘米，根长 3 ~ 4 厘米时便可驯化移栽。第一天打开培养瓶口 1/4 ~ 1/3，第二天打开 1/2，第三天全部打开，逐渐驯化。否则开始揭瓶口太大，叶片易干枯；揭口太小，影响瓶内外气体交换。

一般驯化 3 ~ 5 天后进行移栽。清洗试管苗前 5 分钟将瓶内倒入少许水，软化培养基。将小苗轻轻取出，用自来水清洗干净根部培养基后，移栽于备好的营养钵中。栽苗深度应适宜，下不露根，上不埋苗心。营养钵摆放在炼苗盘中，放在温室或大棚炼苗池中，喷水湿润。

3. 炼苗设施与管理

严格控制炼苗初期的温湿度，炼苗初期（3 ~ 5 天）要求气温昼夜温度在 15 ℃ ~ 30 ℃。温暖季节，加盖 70% ~ 90% 的遮阳网，特别是中午，打开通风口，通风口架设防虫网。必要时可对遮阳网、棚膜喷凉水，降低内部温度，以免引起小苗灼伤、腐烂。叶面酌情喷水，保持空气湿度。5 ~ 7 天后，小苗始发新根，酌情浇水，喷洒营养液、多菌灵等。20 ~ 25 天小苗成活，即可定植。

秋季炼苗，备有保温被或保温草帘，酌情增加增温设施，如红外灯既可增温又可增光。

冬季室内炼苗，宜搭建多层培养架，充分利用空间，架面铺垫塑料布，防止浇水时漫流，架顶布设日光灯，增加光照，光照强度 2 000 ~ 3 000 lx，适当延长光照时间，可弥补室内光照不足。

【注意事项】

（1）珍珠岩必须用水冲洗，以降低 pH，因其相对密度较轻，粉尘污染严重，可在袋内用水淋洗。

（2）清洗试管苗根部培养基时，顺着根毛生长的方向清洗，不要揉洗，用手指捏洗，清洗要彻底。

（3）管理冬季室内炼苗，应将苗整齐摆放在架面塑料布上，用喷头均匀喷水湿润基质，不要将水溅到灯管上，或喷水时关闭灯管电源。

（4）为保证炼苗成活率，其一要培育出健壮的生根苗；其二具备适宜的炼苗基质和能控光、控温、控湿等完善的炼苗设施；其三是炼苗期管理技术，这是保证成活率的关键技术。

【思考题】

（1）试管苗为什么不能直接移植大田？

（2）提高炼苗成活率的基本条件是什么？

XXⅢ. Domestication and Transplanting of Test-tube Seedlings

【Objectives】

(1) Understand and master the growth environment and characteristics of test-tube seedlings.

(2) Master the methods of acclimation, cultivation and transplanting of test-tube seedlings, the preparation of cultivation medium and the management technology during cultivation.

【Principle】

The obtaining and rooting of test-tube seedlings is only a part of tissue culture. Acclimation is an important link between test-tube seedlings and field production seedlings. From inside to outside of the bottle, the test-tube seedlings experiences change from constant temperature, high humidity, weak light and sterile living environment to variable temperature, humidity and bacteria living environment. The changes are very fierce. Moreover, the test-tube seedlings are young, having thin cuticle and weak resistance. Therefore, it is necessary to acclimate and refine seedlings before transplanting in the field. At the initial stage, the air should maintain a certain temperature and humidity, and the substrate should be fertile, ventilated and moist. The survival rate of seedling directly affects the cost of early work and the later field planting.

【Apparatus, materials and reagents】

1. Apparatus and equipment

Seedling refining plate, greenhouse and plastic shed.

2. Materials

Test-tube seedlings, peat soil, perlite, vermiculite, sandy soil and nutrients.

【Procedures】

1. Preparation of the seedling-refining medium

Different plants need different seedling-refining media. It is required that the substrate should be fertile and have good permeability, which is conducive to water permeability and root growth. It is the basic condition to improve the survival rate of seedlings. When prepare the medium, spray disinfectant with 500 – 800 times diluted carbendazim or methyl tozin and other bactericides (a small quantity of matrix can be sterilized in a pressure cooker). Use misty water to mix the substrate. The humidity of the substrate is about 60%. That is to say, if you hold it in a ball, you can fall it to the ground and disperse it. Don't let the humidity be too high, so as to avoid rotten roots and stems. Mix the matrix well and put it in a bowl for standby. The storage time should not exceed half a day.

Generally, peat soil, perlite, vermiculite and sandy soil are used for plants. After disinfection and sterilization, they are mixed according to a certain proportion, which is slightly changed in different seasons. The matrix ratio of common plants:

①Peat soil : perlite : vermiculite = 3 : 1 : 1.

②Peat soil : perlite = 2 : 1 (used to refine seedlings in summer, because vermiculite is easy to keep water and raise temperature).

③Peat soil : sandy soil = 1 : 1.

Orchid class can choose seaweed as seedling-refining medium.

2. Acclimation and transplanting of test-tube seedlings

When the plant grows to 4 – 5 cm in rooting medium and the root length is 3 – 4 cm, it can be domesticated and transplanted. On the first day, open 1/4 – 1/3 of the culture bottle mouth, 1/2 on the second day and all on the third day to gradually domesticate. Otherwise, if the opening is too large, the leaves will dry up easily; if the opening is too small, the gas exchange inside and outside the bottle will be affected.

Generally, the plants will be transplanted after 3 – 5 days of domestication. Pour a little water into the bottle 5 minutes before cleaning the test-tube seedlings and soften the medium. Take out the seedlings gently, clean the root medium with tap water, and transplant them into the prepared nutrition bowl. The depth of seedling should be suitable, the root should not be exposed below, and the seedling core should not be buried above. The nutrition bowl is placed in the seedling refining plate, placed in the greenhouse or seedling-refining pool of shed, and watered.

3. Seedling-refining facilities and management

The temperature and humidity should be strictly controlled at the initial stage of refining seedling. The temperature should be 15 ℃ – 30 ℃ in the daytime and night at the initial stage of refining seedling (3 – 5 d) . In warm season, cover 70% – 90% of the sunshade net, especially at noon, open the vent and set up insect proof net. If necessary, cool water can be sprayed on the sunshade net and shed film to reduce the internal temperature, so as to avoid burning and rotting of seedlings. Spray water on the leaves to keep the air humidity. After 5 – 7 days, the seedlings initiate new roots, water as appropriate, spray with nutrient solution and carbendazim. After 20 – 25 days, the seedlings survive and can be planted.

If seedling in autumn, a heat preservation quilt or a heat preservation straw curtain is required, and heating facilities can be added as appropriate, such as infrared lamp, which can increase both temperature and light.

If conducting indoor refining seedling in winter, it is advisable to build multi-layer culture frame, make full use of space, lay plastic cloth on the scaffold surface to prevent flowing during watering. The roof of the scaffold is equipped with fluorescent lamp to increase the illumination which intensity is 2 000 – 3 000 lx. If the lighting time is extended properly, the indoor illumination shortage can be made up.

【Notes】

(1) Perlite must be washed with water to reduce pH, because of its light relative density and serious dust pollution. It can be washed with water in the bag.

(2) When cleaning the root culture medium of test-tube seedlings, wash along the direction of root hair growth. Do not knead and wash, pinch and wash with fingers, and should wash thoroughly.

(3) For the management of indoor refining seedling in winter, the seedlings shall be placed on the plastic cloth on the shelf surface in order, and the spray nozzle shall be used to spray water evenly to wet the substrate. Do not splash water on the lamp tube, or turn off the power supply of the lamp tube when spraying water.

(4) In order to ensure the survival rate of the seedlings, firstly, we should cultivate strong rooting seedlings; secondly, we should have suitable seedling substrate and perfect seedling facilities that can control light, temperature and humidity; thirdly, we should manage the seedling period, which is the key technology to ensure the survival rate.

【Questions】

(1) Why can't test-tube seedlings be transplanted directly?

(2) What are the basic conditions for improving the survival rate of seedlings?

二十四、植物茎尖分生组织剥离和培养

【实验目的】

掌握茎尖剥取、接种、培养等技术。

【实验原理】

多数栽培植物，尤其是无性繁殖植物，都易受到一种或几种，甚至几十种病毒的侵染。并且随着栽培时间的推移，侵染病毒的种类越来越多。受病毒侵染的植物生长缓慢、畸形、产量大幅度下降，品质变劣，甚至完全丧失商品价值。因此说，病毒带来极大的危害。目前还没有特效药物能治疗病毒引起的疾病。种子一般不带病毒，种子繁殖植物可以得到无病毒植株。但对于无性繁殖植物，必须采用一种有效的方法脱除病毒，使植物恢复高产、优质特性。研究表明，病毒在根上和茎上的分布是不均匀的，离根尖和茎尖越近，病毒的密度越低；反之越高。即病毒在植物体内的分布是不均匀的，分生组织一般无病毒侵染。分生组织之所以能避开病毒侵染，可能有四方面原因：

①在植物体中，病毒的移动主要靠两条途径，一是通过维管系统，而分生组织中尚未形成维管系统；二是通过胞间连丝，但这条途径病毒移动速度非常缓慢，难以追赶上活跃生长的茎尖和根尖。

②旺盛分裂的分生组织代谢活动很高，使病毒无法进行复制。

③在茎尖中存在高水平的内源激素，可以抑制病毒的增殖。

④在植物体内存在有一种“病毒钝化系统”，它在分生组织中的活性最高，因而使分生组织不受侵染。

以上四种可能原因，无论哪种说法正确，事实是分生组织一般不带病毒。因此，利用这一原理可进行茎尖培养，培育无病毒苗。此外，由于茎尖分生组织再分化能力强，也可用于离体快速繁殖。

茎尖培养再生植株途径可通过向培养基中添加不同激素来调控器官发生或胚胎发生。一般在添加适宜较高浓度2，4－D的培养基上，可诱导形成胚性愈伤组织和体细胞胚；在添加适宜浓度生长素（如NAA）和细胞分裂素（如BAP）的培养基上，可诱导形成愈伤组织，进一步分化不定芽，或不经过愈伤组织直接形成丛生芽。

本实验材料甘薯为块根植物，它极易生根，因此无须进行生根培养。

【实验仪器、材料与试剂】

1. 仪器与设备

超净工作台、解剖镜、镊子、解剖针、手术刀、剪刀、酒精灯、棉球、烧杯、火柴、培养皿（无菌）。

2. 培养基

（1）初始培养基：器官发生培养基配方 MS +0.2 mg/L NAA +2.0 mg/L BAP；胚胎发生培养基配方 MS +2，4 - D（0.2 ~2.0 mg/L）。

（2）继代及生根培养基：MS 基本培养基。

以上培养基均添加 3% 蔗糖和 0.8% 琼脂，调 pH 至 5.8。

3. 材料

甘薯茎蔓。

4. 试剂

0.1% 升汞、70% 酒精、无菌水等。

【实验步骤】

1. 取材

薯块育苗，促使其旺盛生长。当出芽后，取 5 厘米左右蔓顶作为实验材料。

2. 材料的消毒

将茎蔓上展开的叶片去掉，用自来水冲洗干净后，在超净工作台内进行表面杀菌。

3. 茎尖剥取与接种

在超净工作台内将材料表面杀菌后，置于经高压灭菌的培养皿上，在解剖镜下用解剖针或刀片层层剥去外面的幼叶和叶原基，剥离至分生组织暴露出来，带 1 ~2 个叶原基，然后从基部切下，随即接种到培养基上，封口。分生组织为透明半圆形。

4. 初始培养

茎尖初始培养基因植株再生途径不同而不同。将茎尖置于器官诱导或体胚诱导培养基上，培养 5 ~7 周后，可分别形成不定芽或体细胞胚。不定芽转移 MS 基本培养基上，茎伸长并长出根成完整植株；体细胞胚转移到 MS 基本培养基上后，萌发长出茎叶和胚根，最终长成完整植株。

5. 大量增殖及生根培养

试管苗可利用茎蔓腋芽继代增殖。将试管苗每 1 ~2 节切成一段，接种到新的 MS 基本培养基上，长大后再切割继代，以达到增殖的目的。由于甘薯容易生根无须用添加激素的培养基诱导生根。

培养条件：27 ℃ ±1 ℃，每日 13 小时，光照 3 000 lx。

6. 驯化移栽

移栽之前，逐步打开瓶盖，使试管苗逐渐适应外界干燥的环境，炼苗 3 ~4 天即可移栽。将小苗从瓶中取出，洗净根部培养基，注意尽量避免损伤茎叶和根。移栽基质选用不灭菌的沙子。移栽后 1 周内保持较高的空气湿度（80% ~90%）。如果条件允许，实验可在驯化室内进行。如果没有驯化室，可盖上塑料纸保持湿度。

【注意事项】

（1）茎尖剥取要用锋利的刀片。

（2）切取茎尖的大小与培养成活率和脱毒效果有直接关系。茎尖越大，成活率越高，但未脱除病毒的可能性也越大。病毒及寄主不同，病毒侵染茎尖的程度不同，而一种植物往往带有多种病毒。为了脱除各种病毒，在保证成活的情况下，茎尖尽量取小为好；从基部切下带 1 ~2 个叶原基的分生组织，避免带有分生组织附近的其他组织。

【思考与记录】

（1）脱毒苗和无菌苗有何区别？

（2）影响茎尖培养的因素有哪些？

（3）填写下列调查表。

不定芽或体细胞胚形成调查表

观察日期　　年　　月　　日

培养基	接种茎尖外植体数	形成不定芽的外植体数	形成芽数	形成体胚的外植体数	形成体胚数

XXIV. Meristem Dissection and Culture of Plant Stem Tip

[Objective]

Master the techniques of stem tip stripping, inoculation and culture.

[Principle]

Most cultivated plants, especially asexual reproduction plants, are vulnerable to one or more, or even dozens of viruses. And with the passage of cultivation time, there are more and more kinds of infected viruses. The plants infected by the virus grow slowly and deform, becoming less productive; its quality becomes inferior, and even lose the commercial value completely. Therefore, the virus brings great harm. At present, there are no special drugs to cure the virus disease. Seeds generally do not carry viruses; seed breeding plants can get virus-free plants. However, for asexual reproduction plants, an effective method must be adopted to remove the virus and restore high yield and high quality characteristics. The results show that the distribution of viruses on the root and stem is uneven. The closer it is to the root tip and stem tip, the lower the density of viruses it is. That is, the distribution of viruses in plants is uneven, meristem is generally free of virus infection. There may be four reasons why meristem can avoid virus infection:

①In plants, virus moves mainly through two pathways, one is through the vascular system, while the vascular system has not yet been formed in the meristem, and the other is through the intercellular filaments, but virus moves very slowly, making it difficult to catch up with the active stem tips and root tips.

②High metabolic activity in vigorously divided meristem prevents replication of the virus.

③The presence of high levels of endogenous hormones in the stem tip inhibits virus proliferation.

④There is a "virus passivation system" in plants, which has the highest activity in meristem, thus leaving meristem free from infection.

These are four possible reasons. No matter which statement is correct, the fact is that meristem is generally free of viruses. Therefore, this principle can be used to cultivate virus-free seedlings. In addition, because of the strong ability of stem apical meristem redifferentiation, it can also be used for rapid propagation *in vitro*.

The regeneration pathway of stem tip cultivation can regulate organogenesis or embryogenesis by adding different hormones to the medium. The embryogenic callus and somatic embryo can be induced by adding medium with suitable high concentration of 2, 4-D, and callus can be induced, further differentiated the adventitious buds or form cluster buds directly without callus by adding medium with suitable concentration of auxin (such as NAA) and cytokinin (such as BAP).

This experiment adopts sweet potato as the root plant, which is easy to take root, so there is no need for rooting culture.

【Apparatus, materials and reagents】

1. Apparatus

Super-clean worktable, anatomic microscope, tweezers, anatomical needle, scalpel, scissors, alcohol lamp, cotton ball, beaker, match, Petri dishes (sterile).

2. Medium

(1) Initial culture medium: organogenesis pathway MS + 0.2 mg/L NAA + 2.0 mg/L BAP; embryogenesis pathway MS + 2, 4-D (0.2 - 2.0 mg/L).

(2) Secondary and rooting medium: MS basic medium.

3% sucrose and 0.8% agar are added to the above medium and adjust pH to 5.8.

3. Materials

Sweet potato stem and vine.

4. Reagents

0.1% mercuric chloride, 70% ethanol and aseptic water, etc.

【Procedures】

1. Materials

Promote potato seedling grow vigorously. After its budding, about 5 cm of vines are taken as experimental materials.

2. Disinfection of materials

Remove the spreading leaves from the stem or vine, rinse with tap water, and sterilize the surface in the super-clean worktable.

3. Stem tip stripping and inoculation

After sterilizing the surface of the material in the super-clean worktable, it is placed on a Petri dish sterilized by high pressure. Under anatomic microscope, the young leaves and leaf primordium are stripped layer by layer with anatomical needle or blade until meristem is exposed. Take 1 – 2 leaf primordium, then cut off from base and inoculate to the medium. Seal it. The meristem is transparent semicircle.

4. Initial culture

The initial medium of stem tip is different due to different regeneration pathways. After 5 – 7 weeks of culture, adventitious buds or somatic embryos could be formed by placing the stem tip on organ induction or somatic embryo induction medium. When adventitious buds are transferred to MS basic medium, stems elongate and develop roots to grow into complete plants; when somatic embryos are transferred to MS basic medium, stems, leaves and radicles germinate and grow into complete plants.

5. Proliferation and rooting culture

A test-tube seedling can be used to proliferate with axillary buds of stem and vine. Cut the seedling into one segment every 1 – 2 sections, inoculate them on a new MS basic medium. Cut them into subculture after the seedlings growing up again. The purpose of proliferation can be achieved by repeating the procedure. Because sweet potato is easy to take root, it does not need to induce root with hormone medium.

Culture conditions: 27 ℃ ± 1 ℃, 13 h per day, daily illumination 3 000 lx.

6. Domestication and transplanting

Prior to transplanting, gradually open the bottle cap, so that the test-tube seedlings gradually adapt to the external dry environment. Seedlings can be transplanted after 3 – 4 d refining. Remove the seedlings from the bottle, wash the root medium, and avoid damaging the stems, leaves and roots. The transplanting medium is unsterilized sand. Maintain high air humidity within 1 week after transplanting (80% – 90%). If conditions allow, the experiement can be carried out in the domestication room. If there is no domestication room, cover plastic paper to maintain humidity.

【Notes】

(1) Use a sharp blade to peel the tip of the stem.

(2) The size of stem tip is directly related to the survival rate and virus-free effect. The larger the stem tip, the higher the survival rate, but the greater the possibility of not removing the virus. The virus and host are different, and the degree of virus infecting stem tip is different. A

plant often has a variety of viruses. In order to remove all kinds of viruses and ensure survival, the stem tip must be as small as possible, and the meristem is cut from the base with 1 – 2 leaf primordia to avoid other tissues near the meristem. .

【Thinking and recording】

(1) What is the difference between virus-free seedlings and sterile seedlings?

(2) What are the factors affecting stem tip culture?

(3) Fill in the following questionnaire.

Questionnaire on adventitious bud formation or somatic embryogenesis

Date of investigation: ______________

Medium	Number of stem apical explants inoculated	Number of explants forming adventitious buds	Number of buds formed	Number of explants of somatic embryos	Number of somatic embryos

二十五、马铃薯茎尖脱毒

【实验目的】

（1）掌握马铃薯茎类培养中去病毒的原理和技术。

（2）了解植物病毒的检测方法。

【实验原理】

马铃薯（*Solanum tuberosum*）是一种全球性的重要作物，在我国分布也很广，种植面积占世界第二位。由于其生长期短，产量高，适应性广，营养丰富，又耐贮藏运输，是一种重要的粮蔬两用作物。马铃薯在种植过程中极易感染病毒，危害马铃薯的病毒有 17 种之多。马铃薯是无性繁殖作物，病毒在母体内增殖、转运和积累于所结的薯块中，并且世代传递，逐年加重，降低其产量和品质。利用茎尖分生组织离体培养技术对已感染的良种进行脱毒处理，获得无病毒的马铃薯植株，对马铃薯增产效果极为显著。

寄生在马铃薯块茎中的病毒，随着块茎芽萌发长成植株的生长过程，也在马铃薯植株体内进行病毒粒子的复制繁殖，但病毒在马铃薯植株内的分布是不均匀的。据研究，在代谢活跃的茎尖分生组织中没有病毒。可能是由于茎尖分生组织中的细胞分裂速度很快，超过病毒粒子的复制速度，使病毒粒子在复制过程中得不到营养而受到抑制，也可能是由于分生组织中某些高浓度的激素抑制了病毒。以上原因的机理尚未厘清，但通过对茎尖（带有 1 ~2 个叶原基，小于 0.2 毫米）组培苗进行病毒检测未发现带有病毒，而大于 0.2 毫米的茎尖却常能检测出病毒。这点便成为茎尖脱毒组培繁殖无病毒株的重要依据。为了提高马铃薯的脱毒效率，可以对外植体材料进行热处理。

【实验仪器、材料与试剂】

1. 仪器

双筒解剖镜、超净工作台、解剖针、解剖刀、镊子、高压灭菌锅、玻璃器皿、锥形瓶等。

2. 材料

马铃薯植株的腋芽和顶芽。

3. 试剂

MS 等组织基本培养基配方所需各种药品、6－BA、NAA、次氯酸钠、酒精、无菌水。

【实验步骤】

1. 取材

在马铃薯生长季节，选取长势旺盛、无明显病害虫害的植株，取其腋芽和顶芽。顶芽的茎尖生长要比取自腋芽的快，成活率也高。为了容易获得无菌的茎尖，应把供试植株种在无菌的盆土中，放在温室进行栽培。对于田间种植的材料，还可以切取插条，在实验室的营养液中生长。由这些插条的腋芽长成的枝条，要比直接取自田间的枝条污染少得多。

2. 灭菌

切取 2～3 厘米的壮芽，去掉所有叶片，用自来水充分洗净，并连续冲洗 0.5～1 小时。然后用 75% 的酒精漂洗 1 分钟，用无菌水冲洗 3 遍，再用 2% 的次氯酸钠消毒 10 分钟，在消毒过程中需要不停地振荡，然后再用无菌水冲洗 3～5 次。置于灭过菌的工具皿上。

3. 脱毒材料的热处理

于 10 倍解剖镜下，用解剖刀切取 1～1.5 mm 的茎尖，将取下的茎尖接种在 MS 培养基上，于 25 ℃每天光照 16 小时的培养室培养。待茎尖长至 1 厘米时转入光照培养箱培养，以每天 16 小时光照，36 ℃的高温处理 6～8 周。

4. 茎尖剥离

在超净工作台上，在 40 倍解剖镜下，用解剖针小心除去茎尖周围的小叶片和叶原基，暴露出顶端圆滑的生长点，用解剖刀细心切取所需的茎尖分生组织，最后只保留带一个叶原基的生长点，大小为 0.1～0.2 毫米。切取的茎尖分生组织随即接种到马铃薯茎尖培养基（MS＋0.5 mg/L NAA＋3% 蔗糖＋0.7% 琼脂，pH 5.8），以切面接触琼脂，封严瓶口置于培养室进行离体培养。

5. 茎尖培养

茎尖培养条件是，温度 23 ℃～25 ℃，光照强度 25～50 μmol/（m^2·s），光照时间为每天 16 小时左右，在正常条件下，经过 30～40 天的培养可见到茎尖有明显的增长，继代 2～3 次，然后将其移入生根培养基（1/2 MS＋1.5% 蔗糖＋0.7% 琼脂，pH 5.8）生根。

6. 脱毒苗的炼苗及移栽

待无菌苗的根长到 1～2 厘米长，将培养瓶上的覆盖物揭掉，在光照培养室内炼苗 1～2 天，在此过程中，要注意保持培养室中的湿度。然后将根部的培养基洗净，移栽到装有蛭石和草炭（1∶2）的培养基中，注意保持空气湿度，1 周就可以成活。

7. 病毒检测

（1）用指示植物检测病毒：检测指示植物病毒一般都用小叶嫁接方法。在嫁接前 1～

2个月，先将生长健壮的指示植物单株栽于盆中，成活后要注意防治蚜虫。从被鉴定植物上取1~3克幼叶，在pH 7.0的磷酸缓冲液中研磨至匀浆，用两层纱布过滤，去渣。在指示植物叶面上涂抹或喷洒滤液，保温15 ℃~25 ℃。接种2~6天后观察有无病毒症状出现。

（2）分子生物学检测病毒：即利用RT-PCR检测病毒的存在。利用已经分离到的病毒及其基因序列合成RT-PCR反应的引物，提取待测马铃薯脱毒植株的RNA，利用反转录酶获得其cDNA，再进行PCR反应，琼脂糖凝胶电泳，观察有无目的片段，有目的条带即脱毒不成功。

（3）血清学检测：即ELISA病毒检测技术，这种方法快速简便，灵敏度高。

【注意事项】

（1）一次消毒的芽不能太多，以免消毒后材料放置时间太长，茎尖发生褐变，影响成活率。

（2）茎尖的大小是影响成苗的直接因素，同时也是影响脱毒效果的重要因素，茎尖越小，成苗率越低而脱毒率越高，因此在茎尖培养中要在保证存活的情况下，尽量剥较小的茎尖进行培养以保证脱毒效果。实验显示，带一个叶原基的茎尖是比较合适的茎尖大小，还应注意的是，茎尖的大小和品种以及发芽状态有关，同一品种顶芽的茎尖较大，则侧芽较单薄。

【思考题】

（1）影响茎尖分生组织培养脱毒效果的因素有哪些？

（2）常用植物病毒检测技术有哪些？

XXV. Virus Elimination of Potato Stem Tips

【Objectives】

(1) Master the principle and technology of virus elimination in potato stem culture.

(2) Understand the detection methods of plant virus.

【Principle】

Potato (*Solanum tuberosum*) is a globally important crop and also widely distributed in China, and its planting area in China occupies the second place in the world. Because of its short growth period, high yield, wide adaptability, rich nutrition, storage and transportation advantages, it is an important dual-use crop as food and vegetable. Potatoes are highly susceptible to viruses during planting, and there are as many as 17 kinds of viruses that harm potatoes. Potato is an asexual breeding crop. The virus multiplies, transports and accumulates in the tubers formed in the mother's body, and is transmitted from generation to generation, increasing year by year, reducing potato's yield and quality. The *in vitro* culture technique of shoot apex meristem is used to eliminate the virus infected seedling to obtain virus-free potato plants, which has a very significant effect on increasing potato production.

The virus parasitic in the potato tuber, along with the tuber budding, germinates and grows into the growth process of the plant. Virus particles also replicate in the potato plant, but the distribution of viruses in the potato plant is uneven. According to research, there is no virus in the metabolically active apical meristem. It may be due to the rapid division of cells in the apical meristem, which exceeds the replication speed of virus particles, preventing virus particles from getting nourishment during the replication process and inhibits viruses. It may also be due to some high concentrations of hormones in the meristem that inhibit viruses. The mechanism of the above reasons has not been clarified. No virus is detected in stem tips (with 1 – 2 leaf primordia, less than 0.2 mm), but virus is often detected in stem tips larger than 0.2 mm. This becomes an important basis for the propagation of virus-free strains in stem tip virus-free tissue culture. In order to improve the viruses-elimination efficiency of potato, the heat treatment of external plant materials can be used.

【Apparatus, materials and reagents】

1. Apparatus

Double-tube anatomical lens, ultra-clean worktable, dissecting needle, scalpel, tweezers, autoclave, Petri dish and Erlenmeyer flasks, etc.

2. Materials

Axillary buds and terminal buds of potato plants.

3. Reagents

MS and other tissue culture media, 6-BA, NAA, sodium hypochlorite, alcohol and sterile water.

【Procedures】

1. Material preparation

During the potato growing season, select plants with vigorous growth and no obvious diseases and pests, pick up the axillary buds and terminal buds. Stem tip of terminal bud grows faster than the axillary bud, and the survival rate is also higher. In order to easily obtain a sterile stem tip, the test plants should be planted in sterile potting soil and placed in a greenhouse for cultivation. For the materials grown in the field, cuttings can also be cut and grown in the nutrient solution in the laboratory. The branches grown from the axillary buds of these cuttings are much less polluted than the branches taken directly from the field.

2. Sterilization

Cut 2 – 3 cm strong buds, remove all leaves, wash thoroughly with tap water, and continuously rinse for 0.5 – 1 h. Then rinse with 75% alcohol for 1 minute, rinse with sterile water for 3 times, and then sterilize with 2% sodium hypochlorite for 10 minutes in order. During the disinfection process, keep shaking. Then rinse with sterile water 3 to 5 times. Place it on a sterilized tool dish.

3. Heat treatment of virus-eliminating materials

Under a 10-fold anatomical len, use a scalpel to cut the stem tip of 1 to 1.5 mm, inoculate the removed stem tip on MS medium, and cultivate it in a culture room at 25 ℃ with 16 hours of light per day. When the stem tip grows to 1 cm, culture it in a light incubator and treat under 16 hours of light per day and a high temperature of 36 ℃ for 6 to 8 weeks.

4. Stem apex peeling

On the ultra-clean worktable, under a 40-fold anatomical lens, carefully remove the small leaves and leaf primordia around the stem tip with a dissecting needle to expose the smooth growth point at the top. Use a scalpel to carefully cut the desired stem apex meristem. Finally, only the growth point with a leaf primordium is retained and the size is 0. 1 to 0. 2 mm. The cut stem apical meristem is then inoculated into potato stem apex medium (MS + 0. 5 mg/L NAA + 3% sucrose + 0. 7% agar, pH 5. 8) with the cut surface touching the agar, and seal the bottle tightly. Perform *in vitro* culture in the culture room.

5. Stem tip culture

The stem tip culture conditions are: 23 ℃ –25 ℃, 25 –50 μmol/ (m^2 · s) of light intensity; light time is about 16 hours per day. Normally, after 30 –40 days of culture, the shoot tip can grow obviously. After subculture for 2 –3 times, the shoot tip should be transferred to rooting medium (1/2 MS + 1. 5% sucrose + 0. 7% agar, pH 5. 8) for rooting.

6. Refining and transplanting of virus-free seedlings

When the roots of the sterile seedlings grow to 1 –2 cm long, remove the covering on the culture flask and cultivate the seedlings in the light culture room for 1 –2 d. During this process, pay attention to maintaining the humidity in the culture room. Then wash the root medium and transplant it to a medium containing vermiculite and peat (1 : 2) . Pay attention to maintaining the air humidity, and it can survive in 1 week.

7. Virus detection

(1) Use indicator plants to detect viruses: Generally, leaflet grafting method is used to detect viruses of indicator plants. 1 to 2 months before grafting, first plant a single strain of a robust indicator plant in a pot, and pay attention to the prevention and control of aphids after survival. Take 1 –3 g young leaves from the identified plants; grind them in a phosphate buffer of pH 7. 0 to homogenize. Filter it with two layers of gauze, and remove the residue. Smear or spray the filtrate on the leaf surface of the indicator plant and keep it at 15 ℃ –25 ℃. Observe whether there are symptoms of viral diseases after inoculation for 2 –6 days.

(2) Molecular biology detection of viruses: RT-PCR is applied to detect the presence of viruses. The RT-PCR reaction primers are synthesized from viruses and their gene sequences that had been isolated, and the RNA of the tested virus-free potato plants is extracted. The cDNA is obtained by reverse transcriptase, and then PCR reaction is performed. Agarose gel electrophoresis is used to observe the target fragments. If there is a target band, the virus is unsuccessfully eliminated.

(3) Serological detection: ELISA virus detection technology. This method is fast, simple and highly sensitive.

【Notes】

(1) There should not be too many buds to be disinfected at a time, so as to avoid that the material being placed for too long time after disinfection, the stem tip will become brown, which will affect the survival rate.

(2) The size of the stem tip is a direct factor that affects seedling formation, and it is also an important factor affecting the virus-eliminating effect. The smaller the stem tip, the lower the seedling rate and the higher the virus-eliminating rate. Therefore, in the stem tip culture, we should peel the smaller shoot tip as much as possible to ensure the virus-free effect and a high survival rate. The experiment shows that the stem tip with a leaf primordium is the more appropriate size of the stem tip. It should also be noted that the size of the stem tip is related to the variety and germination state. The stem tip of the terminal bud of the same variety is larger and the lateral bud is thinner.

【Questions】

(1) What are the factors that affect the virus-eliminating effect of stem apical meristem culture?

(2) What are the commonly used plant virus detection techniques?

二十六、植物原生质体的分离和培养

【实验目的】

掌握用酶解处理植物原生质体的纯化、培养等技术。

【实验原理】

原生质体是指除去细胞壁的裸露的球形细胞。由于没有细胞壁的障碍，可以利用人工的方法诱导原生质体融合，也可以作为遗传转化的受体，从外界摄取 DNA、染色体、细菌、病毒、细胞器和质粒等。原生质体也是分离细胞器以及用于基础理论研究的理想材料。这些技术的成功应用依赖于原生质体植株再生体系的建立。

要进行原生质体培养，首先要去除细胞壁。因为细胞壁的主要成分是纤维素、半纤维素和果胶质，因此酶解液中主要含有纤维素酶和果胶酶（或离析酶）。果胶酶或离析酶能降解植物组织细胞的中胶层，从而达到细胞分离的目的；纤维素酶能降解细胞壁的纤维素，从而使细胞壁解离，获得原生质体。此外，离体原生质体的一个基本属性是渗透破损性。因此，酶解液、原生质体洗涤液及培养基中都需要加入适量的渗透压稳定剂（如甘露醇、山梨醇葡萄糖、蔗糖等)。并且，为防止质膜破坏，提高原生质体的稳定性与活性，酶解液中还需要添加 MES 和 $CaCl_2 \cdot 2H_2O$ 等细胞质膜稳定剂。

原生质体的纯化方法有：①漂浮法，酶解液用相对分子质量较大的蔗糖作为渗透压稳定剂，酶解液经粗过滤后于离心管中低速离心，原生质体将漂浮于表面。缺点是比原生质体小的细胞器杂质等也会漂浮于表面，与原生质体混合一起。②沉降法，酶解液用相对分子质量较小的甘露醇作为渗透压稳定剂，酶解液经粗过滤后于离心管中低速离心，原生质体会沉降于管底。缺点是比原生质体大的组织、细胞团也会沉到底部，与原生质体混合一起。③界面法，用蔗糖作为漂浮剂，放于离心管底部，用甘露醇作为酶解液渗透压稳定剂。将待分离的含原生质体的溶液经粗过滤后，慢慢加于漂浮剂之上，离心，原生质体位于两界面中，比原生质体大的组织等会沉到离心管的最底部，比原生质体小的细胞器等留在酶解液中，界面是纯净的原生质体。

【实验仪器、材料与试剂】

1. 仪器

灭菌锅、超净工作台、摇床、pH 计、电子天平、倒置显微镜、血细胞计数板、过滤器（滤膜为 0.22 ~ 0.25 微米）、三角瓶、离心管、移液管、培养皿、蒸发皿、100 目不锈钢过滤网、Para - film 封口膜、酒精灯、记号笔、火柴、脱脂棉等。

2. 材料

胚性愈伤组织、悬浮细胞、胚状体、叶柄或叶片等（1 克左右）。

3. 试剂

表 26 - 1　酶溶液的组成（A 液，活力较弱，适合容易解离的叶柄等）

组成成分	浓度及 pH	质量/50 mL	质量/100 mL
Cellulase R - 10（纤维素酶）	0.4%	0.2 克	0.4 克
Macerozyme R - 10（离析酶）	0.2%	0.1 克	0.2 克
甘露醇	0.6 mol/L	5.465 克	10.93 克
$CaCl_2 \cdot 2H_2O$	0.5%	0.250 克	0.5 克
MES	5 mmol/L	0.053 5 克	0.107 克
pH	5.8		

表 26 - 2　酶溶液的组成（B 液，活力较强，适合难解离的胚性愈伤组织等）

组成成分	浓度及 pH	质量/50 mL	质量/100 mL
Cellulase R - 10（纤维素酶）	2.0%	1.0 克	2.0 克
Pectolyase Y - 23（果胶酶）	0.1%	0.05 克	0.1 克
甘露醇	0.6 mol/L	5.465 克	10.93 克
$CaCl_2 \cdot 2H_2O$	0.5%	0.250 克	0.5 克
MES	5 mmol/L	0.053 5 克	0.107 克
pH	5.8		

表 26 - 3　蔗糖精制溶液的组成

组成成分	浓度	质量/100 mL
蔗糖	20% ~ 25%	20 ~ 25 克

表 26－4　原生质体洗净液（W_s）的组成

组成成分	浓度及 pH	质量/250 mL	质量/500 mL
$CaCl_2 \cdot 2H_2O$	125.0 mmol/L	4.594 5 克	9.189 克
NaCl	154.0 mmol/L	2.250 克	4.500 克
KCl	5.0 mmol/L	0.093 克	0.187 克
葡萄糖	5.0 mmol/L	0.225 5 克	0.451 克
MES	5.0 mmol/L	0.266 5 克	0.533 克
pH	5.8		

4. 培养基

（1）原生质体液体培养基：初始培养基添加 0.3 ~ 0.6 mol/L 甘露醇或山梨醇。3 ~ 4 周后更换新的培养基，甘露醇或山梨醇浓度减半。再用不含有甘露醇或山梨醇的液体培养基进行培养，培养 3 周左右。添加适量水解酪蛋白、酵母提取液等有机物有利于细胞分裂。

（2）固体培养基：培养基中添加适量激素，促使愈伤组织生长及再生。培养基中添加激素的种类及浓度因植物不同而异，可参照该植物品种细胞及组织培养中激素的使用。

【实验步骤】

1. 酶解处理

（1）在培养皿中加 10 mL 酶溶液（表 26－1，表 26－2）。

（2）胚性愈伤组织或悬浮细胞用刀片切碎；胚状体、叶柄或嫩茎切成薄片；叶片撕掉下表皮，很难撕掉下表皮的叶片切成细条，浸渍于酶溶液中。

（3）培养皿封口后，置于 25 ℃ ~ 27 ℃ 黑暗条件下处理。因材料和酶的种类及浓度不同，处理时间有很大差异，最短 3 ~ 4 小时，最长需 20 小时以上。最后 2 小时置于摇床上，45 转/分钟进行振荡处理，使原生质体充分从组织上游离下来。

2. 原生质体的纯化（界面法）

（1）酶溶液处理结束后，用不锈钢丝网粗过滤，除掉未消化组织。

（2）离心管中加 2 mL 20% ~ 25% 蔗糖精制溶液（表 26－3），再将酶溶液沿离心管壁缓慢加入，使之悬浮于蔗糖溶液之上。

（3）以 350 转/分钟的转速离心 10 分钟，原生质体位于两界面。

（4）用移液管将原生质体收集于新的离心管。

3. 原生质体的洗涤

（1）盛有原生质体的离心管中加 W_s（表 26－4）。

（2）150 转/分钟离心 5 分钟。

（3）用移液管将上清液吸掉，以同样方法用 W_s洗 2 遍，液体培养基洗 1 遍，即可用于培养。

4. 原生质体培养（液体浅层培养法）

在倒置显微镜下用血细胞计数板观察并计数原生质体密度，用液体培养基将原生质体密度调整至 10^4 ~ 10^5个/mL，置于 25 ℃黑暗条件下培养。

在液体培养基中原生质体形成小愈伤组织达直径 1 ~2 毫米时，将其转移到固体培养基上培养，诱导不定芽分化或体细胞胚形成。

5. 培养结果的观察

（1）原生质体的活力测定：凡具有活力的原生质体均呈圆球形，有的可在倒置显微镜下观察到明显的胞质环流运动。可用酚藏花红溶液染色，在显微镜下观察。

（2）细胞壁再生的观察：一般培养 1 ~2 天后新的细胞壁开始再生，体积增大，细胞呈椭圆形。取一滴原生质体培养悬液滴于培养皿底部，其上加一滴 25 % 的蔗糖溶液，在显微镜下观察，再生新壁的细胞会发生质壁分离。

（3）细胞分裂的观察：细胞初始分裂的时间因培养物不同而不同，一般培养 2 ~8 天细胞开始分裂，可用倒置显微镜进行观察。培养 2 周后，调查植板率。

植板率 =（形成细胞团数/细胞总数）×100%

【注意事项】

酶见光容易失去活性，在酶溶液配制以及材料酶解处理过程中都应避光。并且酶在高温下失活，因此用过滤法进行灭菌。

【思考题】

（1）纯化原生质体的方法有哪些？

（2）你认为影响原生质体再生的主要因素是什么？

XXVI. Isolation and Cultivation of Plant Protoplasts

【Objective】

Master the techniques of purification and cultivation of plant protoplasts treated by enzymatic hydrolysis.

【Principle】

Protoplasts refer to naked spherical cells with cell walls removed. Because there is no obstacle of the cell wall, protoplasts can be used to induce protoplast fusion with artificial methods. Protoplasts can also be used as a recipient of genetic transformation to take in DNA, chromosomes, bacteria, viruses, organelles, and plasmids from the outside world. Protoplasts are also ideal materials for separating organelles and for basic theoretical research. The successful application of these technologies depends on the establishment of a protoplast plant regeneration system.

To culture protoplast, the cell wall must be removed first. Because the main components of cell wall are cellulose, hemifibrin and pectin, enzymatic hydrolysates mainly contains cellulase and pectinase (or isolating enzyme). Pectinase or isolating enzyme can degrade the middle glial layer of plant tissue cells to achieve the purpose of cell separation; cellulase can degrade the cellulose of cell wall, thereby dissociating the cell wall and obtaining protoplasts. In addition, one of the basic properties of isolated protoplasts is osmotic damage. Therefore, proper amount of osmotic pressure stabilizer (such as mannitol, sorbitol glucose, sucrose, etc) needs to be added to the enzymatic hydrolysate, protoplast washing solution and culture medium. In addition, in order to prevent the destruction of plasma membrane and improve the stability and activity of protoplasts, it is necessary to add MES, $CaCl_2 \cdot 2H_2O$ and other cell plasma membrane stabilizers to the enzymatic hydrolysate.

Purification methods of protoplasts include: ① Floating method. Sucrose with a relatively large molecular weight is used as an osmotic pressure stabilizer in the enzymatic hydrolysate. After the enzymatic hydrolysate is roughly filtered, it is centrifuged in a centrifuge tube at a low speed, and the protoplasts will float on the surface. The disadvantage is that impurities such as organelles smaller than the protoplasts will also float on the surface and mix with the protoplasts. ②Sedimentation method. Mannitol with a relative small molecular mass is used as an osmotic pressure stabilizer in the enzymatic hydrolysate. After the enzymatic hydrolysate is coarsely filtered, it is centrifuged at a low speed in a centrifuge tube, and the protoplasts will settle at the bottom of the tube. The disadvantage is that tissues and cell clusters larger than the protoplasts will sink to the bottom and mix with the protoplasts. ③Interface method. Sucrose is used as a floating agent and placed at the bottom of the centrifuge tube; mannitol is used as an osmotic pressure stabilizer for the enzymatic hydrolysate. After the enzymatic hydrolysate is coarsely filtered, slowly add it onto the floating agent and then centrifuge. Protoplast is located in the the interface. The tissue larger than protoplast will sink to the bottom of the centrifuge tube, and the organelle smaller than protoplast will stay in the enzyme solution. Pure protoplast will be left in the interface.

【Apparatus, materials and reagents】

1. Apparatus

Sterilization pot, ultra-clean worktable, shaker, pH meter, electronic balance, inverted microscope, hemocytometer, filter (filter membrane is 0.22 – 0.25 μm), Erlenmeyer flask, centrifuge tubes, pipette, Petri dish, evaporating dish, 100 mesh stainless steel filter, Parafilm sealing film, alcohol lamp, marker pen, matches, absorbent cotton, etc.

2. Materials

Embryogenic callus, suspension cells, embryoid body, petiole or leaf (about 1 g), etc.

3. Reagents

Table XXVI – 1 Composition of enzyme solution (Solution A with weak activity, suitable for petioles that are easy to dissociate, etc.)

Composition	Concentration and pH	Weight/50 mL	Weight/100 mL
Cellulase R – 10 (cellulase)	0.4%	0.2 g	0.4 g
Macerozyme R – 10 (isolating enzyme)	0.2%	0.1 g	0.2 g
Mannitol	0.6 mol/L	5.465 g	10.93 g
$CaCl_2 \cdot 2H_2O$	0.5%	0.250 g	0.5 g
MES	5 mmol/L	0.053 5 g	0.107 g
pH	5.8		

Table XXVI - 2 Composition of enzyme solution (Solution B with strong vigor, suitable for difficult-to-dissociate embryogenic callus, etc.)

Composition	Concentration and pH	Weight/50 mL	Weight/100 mL
Cellulase R - 10 (cellulase)	2.0%	1.0 g	2.0 g
Pectolyase Y - 23 (pectinase)	0.1%	0.05 g	0.1 g
Mannitol	0.6 mol/L	5.465 g	10.93 g
$CaCl_2 \cdot 2H_2O$	0.5%	0.250 g	0.5 g
MES	5 mmol/L	0.053 5 g	0.107 g
pH	5.8		

Table XXVI - 3 Composition of sucrose solution

Composition	Concentration	Weight/100 mL
Sucrose	20% - 25%	20 - 25 g

Table XXVI - 4 Composition of protoplast washing solution (W_s)

Composition	Concentration and pH	Weight/250 mL	Weight/500 mL
$CaCl_2 \cdot 2H_2O$	125.0 mmol/L	4.594 5 g	9.189 g
NaCl	154.0 mmol/L	2.250 g	4.500 g
KCl	5.0 mmol/L	0.093 g	0.187 g
Glucose	5.0 mmol/L	0.225 5 g	0.451 g
MES	5.0 mmol/L	0.266 5 g	0.533 g
pH	5.8		

4. Culture medium

(1) Protoplast liquid medium: 0.3 - 0.6 mol/L mannitol or sorbitol is added to the initial medium. After 3 - 4 weeks, the culture medium is changed and the concentration of mannitol or sorbitol is halved. The cells are cultured in liquid medium without mannitol or sorbitol for another 3 weeks. Adding proper amount of casein hydrolysate, yeast extract and other organic substances is conducive to cell division.

(2) Solid medium: the growth and regeneration of callus can be promoted by adding appropriate amount of hormones in the medium. The type and concentration of hormone added in the medium vary with different plants. Referring to the usage of hormone in cell and tissue culture of the plant variety is effective here.

【Procedures】

1. Enzymatic treatment

(1) Add 10 mL of enzyme solution to the petri dish (Table XXVI - 1, Table XXVI - 2) .

(2) Embryogenic callus or suspension cells are chopped with a blade; the embryoid body, petiole or tender stem are cut into thin slices; the leaves are peeled off the epidermis; the leaves difficult to tear off the epidermis are cut into thin strips and immersed them in enzyme solution.

(3) After sealing the petri dish, place it in the dark at 25 ℃ - 27 ℃. Due to the different types and concentrations of materials and enzymes, the processing time varies greatly. The shortest time is 3 - 4 h, and the longest time is more than 20 h. In the last 2 h place it on a shaker, and shake at 45 rpm to make the protoplasts fully free from the tissue.

2. Purification of protoplasts (interface method)

(1) After enzymatic treatment is finished, use a stainless steel wire mesh to coarsely filter, so as to remove undigested tissues.

(2) Add 2 mL of 20% - 25% sucrose solution to the centrifuge tube (Table XXVI - 3), and then slowly add the enzyme solution along the wall of the centrifuge tube to suspend it on the sucrose solution.

(3) Centrifuge at 350 rpm for 10 minutes, and the protoplasts are located at the interface.

(4) Collect the protoplasts in a new centrifuge tube with a pipette.

3. Protoplast washing

(1) Add W_s (Table XXVI - 4) to the centrifuge tube containing the protoplasts.

(2) Centrifuge at 150 rpm for 5 minutes.

(3) Aspirate the supernatant with a pipette, and wash it twice with W_s and once with the liquid medium in the same way, and it can be used for culture.

4. Protoplast culture (liquid shallow culture method)

Observe and count the density of protoplasts with a hemocytometer under an inverted microscope. Adjust the density of protoplasts to 10^4 - 10^5 protoplasts/mL with liquid culture medium, and incubate at 25 ℃ in the dark.

When the protoplasts form small callus in the liquid medium grow to a diameter of 1 - 2 mm, transfer them to a solid medium and incubate to induce adventitious bud differentiation or somatic embryo formation.

5. Observation of culture

(1) Determination of protoplasts activity: All viable protoplasts are spherical, and some can be observed under an inverted microscope with obvious cytoplasmic circulation. It can be

stained with phenosafranine solution and observed under a microscope.

(2) Observation of cell wall regeneration: Generally, new cell walls begin to regenerate after 1 – 2 days of culture, the volume increases, and the cells are oval. Take a drop of the protoplast culture suspension and drop it on the bottom of the petri dish. Add a drop of 25% sucrose solution on it, and observe under the microscope that the cells that regenerate the new wall will undergo plasmolysis.

(3) Observation of cell division: The time of initial cell division varies with different cultures. Generally, cells start to divide after 2 – 8 days of culture, which can be observed with an inverted microscope. After 2 weeks of cultivation, the plate planting rate will be investigated.

Plate planting rate = (number of formed cell clusters/total number of cells) × 100%

【Notes】

Enzymes are prone to lose activity when exposed to light, so light should be avoided during the preparation of the enzyme solution and the enzymatic hydrolysis of materials. And the enzyme will be inactivated at high temperature, so it should be sterilized by filtration.

【Questions】

(1) What are the methods of purifying protoplasts?

(2) What do you think are the main factors affecting the regeneration of protoplasts?

二十七、植物细胞悬浮培养

【实验目的】

学习并掌握植物细胞悬浮培养的方法和技术。

【实验原理】

植物细胞的悬浮培养是指将植物游离的单细胞或细胞团按照一定的细胞密度悬浮在液体培养基中进行培养的方法。将外植体离体培养获得的疏松型的愈伤组织悬浮在液体培养基中，并在摇床上振荡培养一段时间后，可形成分散悬浮培养物，将其调节到适宜密度进行培养。细胞培养需要调节到一定密度是因为细胞能够合成某些对细胞分裂所必需的物质，只有当这些物质的内生浓度达到一个临界值时，细胞才能进行分裂。细胞在培养过程中不断地将这些物质释放到培养基中，直到这些物质在细胞和培养基之间达到平衡才停止释放。细胞密度较高时达到平衡的时间相对较短，细胞密度处于临界密度以下时，达不到这种平衡，细胞不能分裂增殖。

良好的细胞悬浮培养体系应具备以下特征：悬浮培养物分散性良好，细胞团较小，一般在 30 ~ 50 个细胞以下；均一性好，细胞形状和细胞团大小大致相同，悬浮系外观为大小均一的小颗粒，培养基清澈透亮，细胞色泽呈鲜艳的乳白或淡黄色；细胞生长迅速，悬浮细胞的生长量一般 2 ~ 3 天甚至更短时间便可增加一倍。

在液体悬浮培养过程中应注意及时进行细胞继代培养，因为当培养物生长到一定时期将进入分裂的静止期。对于多数悬浮培养物来说，细胞在培养到第 18 ~ 25 天时达到最大的密度，此时应进行第一次继代培养。在继代培养时，应将较大的细胞团块和接种物残渣除去。若从植物器官或组织开始建立细胞悬浮培养体系，就包括愈伤组织的诱导、继代培养、单细胞分离和悬浮培养。目前这项技术已经广泛应用于细胞的形态、生理、遗传、凋亡等研究工作，特别是为基因工程在植物细胞水平上的操作提供了理想的材料和途径。经过转化的植物细胞再经过诱导分化形成植株，即可获得携带有目标基因的个体。

【实验仪器、材料与试剂】

1. 仪器

超净工作台、高压灭菌锅、旋转式摇床、水浴锅、倒置显微镜、电子天平、药匙、称量纸、镊子、酒精灯、火柴、棉球、锥形瓶、移液器、pH 计、恒温培养室、漏斗、不锈钢筛、血球计数板。

2. 材料

花生愈伤组织。

3. 试剂

酚藏花红、荧光素双醋酸酯等。

4. 培养基

MSB_S；MSB_S +10.0 mg/L 2，4－D；MS +10.0 mg/L 2，4－D。

【实验步骤】

1. 愈伤组织的诱导和获得

将花生成熟种子进行表面杀菌后接种于 MSB，置于培养基上培养。以 5～7 天龄胚小叶为外植体，培养在 MSB_S + 10.0 mg/L 2，4－D 培养基上，诱导愈伤组织。

2. 细胞的悬浮培养

（1）在无菌条件下，用镊子将愈伤组织夹取出来，放入含有液体培养基（MS +10.0 mg/L 2，4－D，10～15 mL）的三角瓶中并轻轻夹碎。每瓶接种 1～1.5 克愈伤组织。

（2）将已接种的三角瓶置于旋转式摇床上。在 100 转/分钟，25 ℃～28 ℃条件下，进行振荡培养。

（3）经 6～10 天培养后，向培养瓶中加新鲜培养基 10 mL，必要时可用大口移液管将培养物分装成两瓶，继续培养。可进行第一次继代培养。

（4）悬浮培养物的过滤。按步骤（3）继代培养几代后，培养液中应主要由单细胞和小细胞团（不多于 20 个细胞）组成。若仍含有较大的细胞团，可用适当孔径的金属网筛过滤，再将过滤后的悬浮细胞继续培养。

（5）细胞计数。取一定体积的细胞悬液，稀释 2 倍混匀后，取一滴悬液置入血球计数板上计数。

（6）制作细胞生长曲线。为加深对悬浮培养细胞生长动态的了解，可用以下方法绘制生长曲线图：

①鲜重法：在继代培养的不同时间，取一定体积的悬浮培养物，离心收集后，称量细胞的鲜重，以鲜重为纵坐标，培养时间为横坐标，绘制细胞鲜重生长曲线。

②干重法：可在称量鲜重之后，将细胞进行烘干，再称量干重。以干重为纵坐标，培养时间为横坐标，绘制细胞干重生长曲线。

上述两种方法均需每隔 2 天取样一次，共取 7 次，每个样品重复 3 次，整个实验进行期间不再往培养瓶中换入新鲜培养液。

（7）细胞活力的检查：在培养的不同阶段，吸取一滴细胞悬浮液，放在载玻片上，滴一滴0.1%的酚藏花红溶液（用培养基配制）染色，在显微镜下观察。凡活细胞均不着色，而死细胞则很快被染成红色。也可用0.1%荧光素双醋酸酯溶液染色，凡活细胞将在紫外光诱发下显示蓝绿色荧光，也可根据细胞形态、胞质环流判别细胞的死活。

（8）细胞再生能力的鉴定。将培养细胞转移到琼脂固化的培养基上，使其再形成愈伤组织，进而再分化培养基土，诱导植株的分化。

【注意事项】

（1）上述步骤均需无菌操作，培养基、用具、器皿等要高压灭菌后方可使用。

（2）如果培养液混浊或呈现乳白色，表明已污染。

（3）每次继代培养时，应在倒置显微镜下观察培养物中各类细胞及其他残余物的情况以有意识地留下圆细胞，弃去长细胞。

【思考题】

（1）良好的细胞悬浮培养体系应具备什么特征？

（2）怎样检测细胞的活力？

XXVII. Plant Cells Suspension Culture

【Objective】

Learn and master the methods and techniques of plant cell suspension culture.

【Principle】

Suspension culture of plant cells refers to a method in which free single cells or cell clusters of plants are suspended in a liquid medium at a certain cell density for cultivation. The loose callus obtained by explant *in vitro* culture is suspended in a liquid medium. After shaking culture in a shaker for a period of time, a dispersed suspension culture can be formed. The dispersed suspension culture can be cultivated after it is adjusted to a suitable density. It needs to be adjusted to a certain density for cell culture, because cells can synthesize certain substances necessary for cell division. Only when the endogenous concentration of these substances reaches a critical value, cells can divide. In the process of cell incubation, these substances are continuously released into the medium, and the release of these substances will not stop until the balance between the cells and the medium is reached. When the cell density is high, the time to reach equilibrium is relatively short. When the cell density is below the critical density, this equilibrium cannot be reached, and the cells cannot divide and proliferate.

A good cell suspension culture system should have the following characteristics: the suspension culture has good dispersion, and the cell clusters are small, generally less than 30 to 50 cells; the uniformity is good, the cell shape and the cell cluster size are about the same; the suspension system contains particles in same size. The medium is clear and bright, and the cell color is bright milky white or light yellow; the cells grow rapidly, and the growth of suspended cells can generally double in 2 to 3 days or even less.

In the process of liquid suspension culture, attention should be paid to cell subculture in time, because when the culture grows to a certain period of time, it will enter the quiescent phase of division. For most suspension cultures, the cells reach their maximum density at the 18th to 25th day of incubation, and the first subculture should be carried out at this time. During subculture, larger cell clumps and inoculum residues should be removed. If a cell suspension culture system is established from plant organs or tissues, it includes callus induction, subculture, single cell separation and suspension culture. At present, this technology has been widely used in cell morphology, physiology, genetics, apoptosis and other research work, especially for the operation of genetic engineering at the plant cell level to provide ideal materials and methods. The transformed plant cells are then induced to differentiate to form plants, and individuals carrying the target gene can be obtained.

【Apparatus, materials and reagents】

1. Apparatus

Ultra-clean worktable, autoclave, rotary shaker, water bath, inverted microscope, electronic balance, spoon, weighing paper, tweezers, alcohol lamp, matches, cotton ball, Erlenmeyer flasks, pipettes, pH meter, constant temperature incubation room, funnels, stainless steel sieve, hemocytometer.

2. Materials

Peanut callus.

3. Reagents

Phenosafranine and fluorescent diacetate.

4. Medium

MSB_S; MSB_S + 10.0 mg/L 2, 4-D; MS + 10.0 mg/L 2, 4-D.

【Procedures】

1. Induction and acquisition of callus

The mature seeds of peanuts are sterilized on the surface and then inoculated into MSB and cultivated on the medium. The 5 – 7 day-old embryo leaflets are used as explants and cultured on MSBs + 10.0 mg/L 2, 4-D medium to induce callus.

2. Cell suspension culture

(1) Under aseptic conditions, take out the callus clamp with tweezers, put it in an Erlenmeyer flask containing liquid medium (MS + 10.0 mg/L 2, 4-D, 10 – 15 mL) and gently clamp it broken. Each flask is inoculated with 1 – 1.5 g callus.

(2) Place the inoculated Erlenmeyer flask on a rotary shaker. Shake the flask at 100 rpm and 25 ℃ – 28 ℃.

(3) After 6 – 10 days of culture, add 10 mL of fresh medium to the culture flask. If neces-

sary, divide the culture into two flasks with a wide-mouth pipette and continue to culture. The first subculture can be carried out.

(4) Filtration of suspension culture. After several generations of subculture according to step (3), the culture medium should mainly consist of single cells and small cell clusters (not more than 20 cells). If it still contains larger cell clusters, it can be filtered with a metal mesh sieve of appropriate pore size, and then the filtered suspension cells are continued to be cultured.

(5) Cell count. Take a certain volume of cell suspension, dilute it by 2 times and mix it well. Then take a drop of the suspension and place it on a hemocytometer for counting.

(6) Making cell growth curve. In order to deepen the understanding of the growth dynamics of suspension culture cells, the following methods can be used to draw growth curve diagrams:

①Fresh weight method. Take a certain volume of suspension culture at different times of subculture. After centrifugation to collect, weigh the fresh weight of the cells. Use the fresh weight as the ordinate and the culture time as the abscissa to plot the fresh weight growth of the cells curve.

②Dry weight method. After weighing the fresh weight, the cells can be dried, and then the dry weight can be weighed. With the dry weight as the ordinate and the culture time as the abscissa, draw the cell dry weight growth curve.

The above two methods require sampling every 2 days, a total of 7 times, repeat steps on each sample for 3 times, and no fresh culture medium is changed into the culture flask during the entire experiment.

(7) Inspection of cell viability. At different stages of culture, draw a drop of cell suspension; place it on a glass slide. Drop a drop of 0.1% phenosafranine solution (prepared with culture medium) for staining, and observe under a microscope. Live cells will not be stained, while dead cells will be quickly stained red. It can also be stained with 0.1% fluorescent diacetate solution, where live cells will show blue-green fluorescence under UV light induction. Cell morphology and cytoplasmic circulation can also be used to determine whether cells are alive or dead.

(8) Identification of cell regeneration ability. Transfer the cultured cells to agar-solidified medium to re-form callus, and then induce plant differentiation on the differentiation medium soil.

【Notes】

(1) The above steps require aseptic operation, and the culture medium, utensils, containers, etc., must be autoclaved before use.

(2) If the culture medium is turbid or milky white, it indicates that it has been contaminated.

(3) During each subculture, observe the condition of various cells and other residues in the

culture under an inverted microscope to intentionally leave round cells and discard long cells.

【Questions】

(1) What are the characteristics of a good cell suspension culture system?

(2) How to determine the viability of cells?

二十八、动物细胞的原代培养

【实验目的】

学习并掌握原代细胞培养的一般步骤，熟悉原代培养细胞的观察方法。

【实验原理】

细胞培养是生物学研究最常用的手段之一，可分为原代培养和传代培养两种。原代培养是直接从生物体获取细胞进行培养。因细胞刚刚从活体组织分离出来，故更接近于生物体内的生活状态。这一方法可为研究生物体细胞的生长、代谢、繁殖提供有力的手段，同时也为以后传代培养创造条件。

【实验仪器、材料与试剂】

1. 仪器

CO_2培养箱、倒置显微镜、超净工作台、磁力搅拌器、离心机、离心管、血球计数板、水浴锅、玻璃漏斗、三角烧瓶、平皿、试管、移液管、无菌纱布、无菌眼科剪、手术刀、医用镊子、手术器械、大头针、废液缸、蜡盘、电子天平、药匙、称量纸。

2. 实验材料

胎鼠或新生鼠。

3. 试剂

0.25%胰蛋白酶溶液、D－Hank's 液、碘酒、酒精、新洁尔灭。

4. 培养基

（1）1 640 培养基（含 10% 小牛血清）。

（2）Hank's 液配方：取 $CaCl_2$ 0.14 克，$MgCl_2 \cdot 6H_2O$ 0.1 克，KCl 0.4 克，KH_2PO_4 0.06 克，$MgSO_4 \cdot 7H_2O$ 0.1 克，NaCl 8.0 克，$NaHCO_3$ 0.35 克，葡萄糖 1.0 克，$Na_2HPO_4 \cdot 7H_2O$ 0.09 克，酚红 0.02 克，加水至 1 000 mL。

（3）D－Hank's 液配方：KH_2PO_4 0.06 克，NaCl 8.0 克，$NaHCO_3$ 0.35 克，KCl 0.4 克，葡萄糖 1.0 克，$Na_2HPO_4 \cdot H_2O$ 0.06 克，酚红 0.02 克，加水至 1 000 mL。D－Hank's 液可以高压灭菌。4 ℃下保存。

（4）0.25%胰蛋白酶溶液：称取0.25克胰蛋白酶（活力为1：250），加入100 mL无Ca^{2+}、Mg^{2+}的D－Hank's液溶解，过滤除菌，4 ℃保存，用前可在37 ℃下回温。胰蛋白酶溶液中也可加入EDTA，使最终浓度达0.02%。

【实验步骤】

1. 胰蛋白酶溶液消化法

（1）取材：将孕鼠或新生小鼠引颈处死，置75%酒精泡2～3秒（时间不能过长，以免酒精从口和肛门浸入体内），再用碘酒消毒腹部，取胎鼠带入超净工作台内（或将新生小鼠放在超净工作台内）解剖取出肝脏，置平皿中。

（2）用Hank's液洗涤三次，并剔除脂肪、结缔组织、血液等杂物。

（3）用手术剪将肝脏剪成小块（1 mm^2），再用Hank's液洗三次，转移至离心管中。

（4）视组织块量加入5～10倍的0.25%胰蛋白酶溶液，37 ℃水浴中消化20～40分钟，每隔5分钟振荡一次，使细胞分离。

（5）待组织变得疏松，颜色略微发白时，加入3～5 mL培养液（含10 %小牛血清）以终止胰蛋白酶消化作用（或加入胰蛋白酶抑制剂）。

（6）将离心管放入离心机中，1 000转/分钟，离心10分钟，弃上清液。

（7）在离心管中加入Hank's液5 mL，冲散细胞，再离心一次，弃上清液。

（8）加入培养液1～2 mL（视细胞量），血球计数板计数。

（9）将细胞数调至5×10^5个/mL左右，转移至25 mL细胞培养瓶中，37 ℃下培养。

上述消化分离的方法是最基本的方法，在该方法的基础上，可进一步分离不同细胞。细胞分离的方法各实验室不同，所采用的消化酶也不相同（如胶原酶、透明质酶等）。

2. 组织块直接培养法

自上述方法第（3）步后，将组织块转移到培养瓶，贴附于瓶底面。翻转瓶底朝上，将培养液加至瓶中，培养液勿接触组织块，置37 ℃培养箱静置3～5小时，轻轻翻转培养瓶，使组织浸入培养液中（勿使组织漂起），37 ℃继续培养。

【注意事项】

（1）自取材开始，保持所有组织细胞处于无菌条件。细胞计数可在有菌环境中进行。

（2）在超净工作台中，组织、细胞、培养液等不能暴露过久，以避免溶液蒸发。

（3）凡在超净工作台外操作的步骤，各器皿需用盖子或橡皮塞，以防止细菌落入。

（4）操作前要洗手，进入超净工作台后手要用75% 酒精或0.2% 新洁尔灭擦拭。试剂等瓶口也要擦拭。

（5）点燃酒精灯，操作在火焰附近进行，耐热物品要经常在火焰上烧灼。金属器械烧灼时间不能太长，退火并冷却后才能夹取组织。吸取过营养液的用具不能再烧灼，以免烧焦形成碳膜。

（6）操作动作要准确敏捷，但又不能太快，以防空气流动，增加污染机会。

（7）不能用手触已消毒器皿的工作部分，工作台面上用品要布局合理。

（8）瓶子开口后要尽量保持45度斜位。

（9）吸溶液的吸管等不能混用。

【思考题】

（1）胰蛋白酶的作用是什么呢？由此可说明细胞间的物质是什么成分？胰蛋白酶能将细胞消化掉吗？

（2）将动物组织分散成单个细胞要注意什么问题呢？

XXVIII. Primary Culture of Animal Cells

【Objective】

Learn and master the general steps of primary culture and be familiar with the observation methods of primary culture cells.

【Principle】

Cell culture is one of the most commonly used methods in biological research, and it can be divided into primary culture and subculture. Primary culture is to obtain cells directly from organisms for cultivation. Since cells have just been separated from living tissues, they are closer to living conditions in the organism. This method provides a powerful means for research of the growth, metabolism, and reproduction of cells, and at the same time create conditions for subsequent subculture.

【Apparatus, materials and reagents】

1. Apparatus

CO_2 incubator, inverted microscope, ultra-clean worktable, magnetic stirrer, centrifuge, centrifuge tubes, hemocytometer, water bath, glass funnel, triangular flask, plate, test tube, pipettes, sterile gauze, sterile ophthalmic scissors, scalpel, forceps, surgical instruments, pins, waste liquid tank, wax tray, electronic balance, spoon, weighing paper.

2. Materials

Fetal or newborn rats.

3. Reagents

0.25% trypsin solution, D-Hank's solution, iodine, alcohol, bromogeramine.

4. Medium

(1) 1 640 medium (containing 10% calf serum).

(2) Hank's solution formula: 0.14 g $CaCl_2$, 0.1 g $MgCl_2 \cdot 6H_2O$, 0.4 g KCl, 0.06 g KH_2PO_4, 0.1 g $MgSO_4 \cdot 7H_2O$, 8.0 g NaCl, 0.35 g $NaHCO_3$, 1.0 g glucose, 0.09 g

$Na_2HPO_4 \cdot 7H_2O$, 0.02 g phenol red, and add water to 1 000 mL.

(3) D-Hank's solution formula: 0.06 g KH_2PO_4, 8.0 g NaCl, 0.35 g $NaHCO_3$, 0.4 g KCl, 1.0 g glucose, 0.06 g $Na_2HPO_4 \cdot H_2O$, 0.02 g phenol red, and add water to 1 000 mL. D-hank's solution can be sterilized with autoclave. Store at 4 ℃.

(4) 0.25% trypsin solution: weigh 0.25 g trypsin (the activity is 1 : 250), add 100 mL of D-Hank's solution without Ca^{2+}, Mg^{2+}, dissolve, filter and remove bacteria and store at 4 ℃. Return to temperature at 37 ℃ before use. EDTA can also be added to trypsin solution, and the final concentration is 0.02%.

【Procedures】

1. Trypsin solution digestion

(1) Preparation of tissue. The pregnant mouse or newborn mouse is put to death by neck drawing. Put in 75% ethanol for 2 to 3 s (the time should not be too long, so as to avoid alcohol soaking in the body from the mouth and anus), then disinfect the abdomen with iodine, take the fetal mice and put them in the ultra-clean worktable (or put the newborn mice in the ultra-clean worktable) to dissect and take out the liver, and put them in a plate.

(2) Wash them with Hank's solution for 3 times, and remove the impurities such as fat, connective tissue and blood.

(3) The liver should be cut into small pieces (1 mm^2) with surgical scissors. Wash them with Hank's solution for 3 times, and transfer them to the centrifuge tube.

(4) Add 5 to 10 times of 0.25% trypsin solution according to the tissue mass, digest them in water bath at 37 ℃ for 20 to 40 mins, and vibrate every 5 mins to separate the cells.

(5) When the tissue becomes loose and slightly white, add 3 to 5 mL of medium (containing 10% calf serum) to stop trypsin digestion (or add trypsin inhibitor).

(6) Put the centrifuge tube into the centrifuge, 1 000 rpm, centrifuge for 10 mins, and discard the supernatant.

(7) Add 5 mL of Hank's solution into the centrifuge tube to disperse the cells, and centrifuge again to discard the supernatant.

(8) Add 1 to 2 mL of medium (depending on the amount of cells) and count with hemocytometer.

(9) Adjust the number of cells to about 5×10^5 cells/mL, transfer them to a 25 mL cell culture bottle and culture them at 37 ℃.

The above digestion and separation method is the most basic method. On the basis of this method, different cells can be further separated. The methods of cell separation are different in different laboratories, and the digestive enzymes used are also different (such as collagenase, hyaluronidase, etc.).

2. Direct tissue culture

After the third step of the above method, transfer the tissue lump to the culture bottle and attached it to the bottom of the bottle. Invert the bottle, add the medium into the bottle, keep the medium away from the tissue lump, put it in the 37 ℃ incubator for 3 to 5 h, gently turn the culture bottle, immerse the tissue in the medium (do not float the tissue), and continue to culture at 37 ℃.

【Notes】

(1) Keep all tissue cells in sterile condition from the beginning of sampling. Cell counting can be performed in the presence of bacteria.

(2) On the ultra-clean worktable, tissue, cells and medium should not be exposed too long to avoid evaporation of the solution.

(3) Concerning steps taken outside the ultra-clean worktable, each vessel should be plugged with a cap or rubber stopper to prevent bacteria from falling into.

(4) Wash hands before operation, and wipe with 75% ethanol or 0.2% bromogeramine after entering the ultra-clean worktable. Reagents and other bottle mouth should also be wiped.

(5) Light the alcohol lamp and operate it near the flame. Heat-resistant articles should be burned on the flame frequently. The burning time of metal instruments should not be too long, so as to avoid holding the tissue lump after annealing and cooling. The utensils that have absorbed the nutrient solution should not be burned again to avoid the formation of carbon film.

(6) The operation should be accurate and quick, but not too fast, in order to prevent air flowing and the chance of pollution increasing.

(7) Don't touch the working part of sterilized utensils with hands. The articles on the worktable should be arranged reasonably.

(8) Try to keep 45° oblique position after the bottle is opened.

(9) The pipettes for absorbing different solutions should not be mixed.

【Questions】

(1) What is the function of trypsin? What is the composition of intercellular substances? Can trypsin digest cells?

(2) What should we pay attention to when we disperse animal tissues into single cells?

二十九、动物细胞的传代培养

【实验目的】

掌握细胞传代培养的技术。

【实验原理】

细胞在培养瓶长成致密单层后，已基本上饱和，为使细胞能继续生长，同时也将细胞数量扩大，需进行传代再培养。传代培养也是一种将细胞种保存下去的方法，同时也是利用培养细胞进行各种实验的必经过程。悬浮细胞直接分瓶就可以，而贴壁细胞需经消化后才能分瓶。

【实验仪器、材料与试剂】

1. 仪器

CO_2培养箱、倒置显微镜、超净工作台。

2. 材料

小鼠贴壁细胞株。

3. 试剂

0.25%胰蛋白酶溶液、75%酒精。

4. 培养基

1 640 培养基（含 10%小牛血清）。

【实验步骤】

（1）将长满细胞的培养瓶中原来的培养液弃去。

（2）加入 0.5 ~ 1.0 mL 0.25%胰蛋白酶溶液，使瓶底细胞都浸入溶液中。

（3）瓶口塞好橡皮塞，放在倒置显微镜下观察细胞。随着时间的推移，原贴壁的细胞逐渐趋于圆形，在还未漂起时将胰蛋白酶弃去，加入 10 mL 培养液终止消化（观察消化也可以用肉眼，当见到瓶底发白并出现细针孔空隙时终止消化。一般室温消化时间约为 1 ~ 3 分钟）。

（4）用吸管将贴壁的细胞吹打成悬液，分到另外两到三瓶中，置 37 ℃下继续培养。第二天观察贴壁生长情况。

【注意事项】

（1）传代培养时要注意严格的无菌操作，并防止细胞之间的交叉污染。

（2）酶解消化过程中要不断观察，消化过度会对细胞造成损害，消化不够则难于将细胞解离下来。

（3）传代后每天观察细胞生长情况，了解细胞是否健康生长：健康细胞的形态饱满，折光性好。

（4）掌握好传代时机：健康生长的细胞生长致密，即将铺满瓶底时，即可传代。

【思考题】

（1）培养瓶中的细胞培养到什么时候就停止生长和增殖?

（2）如何理解原代细胞和传代细胞?

XXIX. Subculture of Animal Cells

【Objectives】

Master the technology of cell subculture.

【Principle】

After the cells grow into a dense monolayer, they will be basically saturated. In order to make the cells continue to grow and expand the number of cells, it is necessary to subculture. Subculture is a way to preserve cell species, and it is also a necessary process to carry out various experiments with cultured cells. Suspension cells can be directly divided into flasks, while adherent cells can only be divided into flasks after digestion.

【Apparatus, materials and reagents】

1. Apparatus

CO_2 incubator, inverted microscope and ultra-clean worktable.

2. Materials

Adherent mouse cell line.

3. Reagents

0.25% trypsin solution, 75% ethanol.

4. Culture medium

1 640 medium (containing 10% calf serum).

【Procedures】

(1) Discard the original culture medium from the culture bottle full of cells.

(2) Add 0.5 – 1.0 mL 0.25% trypsin solution to immerse the cells at the bottom of the bottle.

(3) Put rubber stopper in the bottle mouth and observe the cell under the inverted microscope. Over time, the cells on the original wall gradually become round. Discard the trypsin before cells are floated, add 10 mL of culture medium to stop digestion (the digestion can also be ob-

served by naked eye, when the bottom of bottle is white and there is a fine pinhole space, stop the digestion. The digestion time at room temperature is about 1 – 3 mins) .

(4) Blow the adherent cells into suspension with suction tube, divide them into 2 or 3 bottles, and then culture them at 37 ℃. Observe the growth situation of adherent cells the next day.

【Notes】

(1) Keep strict aseptic operation during subculture process, and cross contamination between cells should be prevented.

(2) In the process of enzymolysis and digestion, we should constantly observe because excessive digestion will cause damage to the cells, while insufficient digestion will make it difficult to dissociate the cells.

(3) Observe the growth of cells every day after subculture. The standard to know whether the cells grow healthily: the shape of healthy cells is full and the refraction is good.

(4) Grasp the subculture time: the healthy cells can be passaged when they grow compactly and the bottom of the bottle is about to be covered.

【Questions】

(1) When should we stop cell growing and proliferating after we observe certain conditions of cell growth?

(2) How to understand the primary and the subculture cells?

三十、动物细胞的冻存与复苏

【实验目的】

（1）学习并掌握细胞冻存的方法。

（2）能熟练进行细胞冻存与复苏操作。

【实验原理】

在不加任何条件下直接冻存细胞时，细胞内、外环境中的水都会形成冰晶，能导致细胞内发生机械损伤、电解质升高、渗透压改变、脱水、pH 改变、蛋白变性等，从而引起细胞死亡。如向培养液加入保护剂，可使冰点降低。在缓慢的冻结条件下，能使细胞内水分在冻结前透出细胞。贮存在 -130 ℃以下的低温中能减少冰晶的形成。

细胞复苏时速度要快，使之迅速通过细胞最易受损的 -5 ℃ ~0 ℃，细胞仍能生长，活力受损不大。

目前常用的保护剂为二甲基亚砜（DMSO）和甘油，它们对细胞无毒性，相对分子质量小，溶解度大，易穿透细胞。

【实验仪器、材料与试剂】

1. 仪器

4 ℃冰箱、-70 ℃冰箱、液氮罐、离心机、水浴锅、微量加样器、电子天平、药匙、称量纸、血细胞计数板。

2. 材料

小鼠细胞悬液。

3. 试剂

（1）0.25% 胰蛋白酶溶液；

（2）含保护剂的培养基（即冻存液）。

4. 培养基

（1）1 640 培养基（含 10% 小牛血清）。

（2）冻存液配制：培养基加入甘油或 DSMO，使其最终浓度达 5% ~20%。保护剂的种类和用量随细胞不同而有差异。配好后4 ℃下保存。

【实验步骤】

1. 冻存

（1）消化细胞，将细胞悬液收集至离心管中。

（2）将离心管放入离心机中，1 000 转/分钟离心 10 分钟，弃上清液。

（3）将沉淀加入含保护液的培养基，血细胞计数板计数调整至 5×10^6个/mL 左右。

（4）将细胞悬液分至冻存管中，每管 1 mL。

（5）将冻存管口封严。若用安瓿瓶需火焰封口，封口一定要严，否则复苏时易出现爆裂。

（6）贴上标签，写明细胞种类、冻存日期。冻存管外拴一金属重物和一细绳。

（7）按下列顺序降温：室温→4 ℃（20 分钟）→冰箱冷冻室（30 分钟）→低温冰箱（－30 ℃，1 小时）→气态氮（30 分钟）→液氮。

注意：操作时应小心，以免被液氮冻伤。液氮定期检查，随时补充，绝对不能挥发干净，一般 30 立升的液氮能用 1～1.5 个月。

2. 复苏

（1）从液氮中取出冻存管，迅速置于 37 ℃温水中并不断搅动。使冻存管中的冻存物在 1 分钟之内融化。

（2）打开冻存管，将细胞悬液吸到离心管中。

（3）1 000 转/分钟离心 10 分钟，弃去上清液。

（4）沉淀加 10 mL 培养液，吹打均匀，再离心 10 分钟，弃上清液。

（5）加适当培养基后将细胞转移至培养瓶中，37 ℃培养，第二天观察生长情况。

【思考题】

（1）细胞冻存与复苏的基本原则是什么？

（2）冻存液的作用是什么？

XXX. Cryopreservation and Resuscitation of Animal Cells

【Objectives】

(1) Learn and master the method of cell cryopreservation.

(2) Be able to perform the operation of cell cryopreservation and resuscitation.

【Principle】

When cells are cryopreserved directly without any preconditions, the water inside and outside the cell will form ice crystals, which can lead to mechanical damage, electrolyte rise, osmotic pressure change, dehydration, pH change, protein denaturation, etc., leading to cell death. The freezing point can be decreased by adding protective agent to the medium. Under the condition of slow freezing, the water in the cell can penetrate into the cell before freezing. Storage at low temperature below −130 ℃ can reduce the formation of ice crystals.

The speed of cell resuscitation should be fast, so that the cells can quickly pass through the most vulnerable temperature range of −5 ℃ − 0 ℃, the cells can still grow, and the vitality is not damaged.

At present, dimethyl sulphoxide (DMSO) and glycerin are commonly used as protectants. They are non-toxic, low molecular weight, high solubility, easy to penetrate cells and hemocyytometer.

【Apparatus, materials and reagents】

1. Apparatus

4 ℃ refrigerator, −70 ℃ refrigerator, liquid nitrogen tank, centrifuge, water bath, micro sampler, electronic balance, spoon, weighing paper and hemocytometer.

2. Materials

Mouse cell suspension.

3. Reagents

(1) 0.25% trypsin solution.

(2) Medium containing protective agent (i. e. cryopreservation solution).

4. Culture medium

(1) 1 640 medium (containing 10% calf serum).

(2) Preparation of cryopreservation solution: glycerin or DMSO is added to the medium to make its final concentration reach 5% – 20%. The types and dosage of protectants vary with different cells. After preparation, it should be stored at 4 ℃.

【Procedures】

1. Cryopreservation

(1) Digest the cells, and collect the cell suspension into the centrifuge tube.

(2) Put the centrifuge tube into the centrifuge, centrifuge at 1 000 rpm for 10 mins, and discard the supernatant.

(3) Add the precipitate into the medium containing protective solution, and adjust the hemocytometer to about 5×10^6 cells/mL.

(4) Divide the cell suspension into cryopreservation tubes, 1 mL in each tube.

(5) Seal the cryopreservation tube tightly. If ampoules need to be sealed with flame, the sealing must be strict; otherwise it is easy to burst during resuscitation.

(6) Label the cell type and cryopreservation date. A metal weight and a string should be tied outside the cryopreservation tube.

(7) Cooling in the following order: room temperature → 4 ℃ (20 mins) → refrigerator freezer (30 mins) → low temperature refrigerator (–30 ℃, 1 h) → gaseous nitrogen (30 mins) → liquid nitrogen.

Attention: be careful during operation to avoid being frostbitten by liquid nitrogen. Liquid nitrogen should be checked regularly and replenished at any time. It must not be volatilized. Generally, 30 liters of liquid nitrogen can be used for 1 – 1.5 months.

2. Resuscitation

(1) Take out the cryopreservation tube from liquid nitrogen, quickly put it in 37 ℃ warm water and stir it continuously. Thaw the cryopreservation in 1 min.

(2) Open the cryopreservation tube and suck the cell suspension into the centrifuge tube.

(3) Centrifuge at 1 000 rpm for 10 mins, and discard the supernatant.

(4) The precipitate is added to 10 mL of medium, blow evenly, centrifuge for 10 mins, and discard the supernatant.

(5) After adding appropriate medium, transfer the cells to the culture bottle, culture them at 37 ℃, and observe the growth at next day.

【Questions】

(1) What are the basic principles of cell cryopreservation and resuscitation?

(2) What is the function of cryopreservation solution?

三十一、动物细胞的融合

【实验目的】

（1）理解 PEG 诱导细胞融合的基本原理。

（2）通过 PEG 诱导的鸡红血球细胞融合实验，掌握细胞融合技术。

【实验原理】

在诱导剂（如仙台病毒、聚乙二醇等）作用下，相互靠近的细胞发生凝集，随后在质膜接触处发生质膜成分的一系列变化，主要是某些化学键的断裂与重排，进而细胞质交融，形成一个大的双核或多核细胞（此时称同核体或异核体）。

【实验仪器、材料与试剂】

1. 仪器

显微镜、离心机、天平、药匙、称量纸、注射器、血细胞计数器。

2. 材料

年龄为 1 的公鸡静脉血。

3. 试剂

（1）0.85%生理盐水、双蒸水和 Janus Green 染液。

（2）Alsever 溶液：葡萄糖 2.05 克，柠檬酸钠 0.80 克，NaCl 0.42 克，溶于 100 mL 双蒸水中。

（3）GKN 溶液：NaCl 8.0 克，KCl 0.4 克，$Na_2HPO_4 \cdot 2H_2O$ 1.77 克，$NaH_2PO_4 \cdot H_2O$ 0.69 克，葡萄糖 2 克，酚红 0.01 克，溶于 1 000 mL 双蒸水中。

（4）50% PEG 溶液：称取一定量的 PEG（Mr = 4 000）放入烧杯中，沸水浴加热，使之溶解，待冷却至 50 ℃时，加入等体积预热至 50 ℃的 GKN 溶液，混匀，置 37 ℃备用。

【实验步骤】

（1）在公鸡翼下静脉抽取 2 mL 鸡血，加入 8 mL 的 Alsever 溶液中，使血液与 Alsever 溶液的比例达 1∶4，混匀后可在冰箱中存放一周。

（2）取此贮存鸡血 1 mL 加入 4 mL 0. 85% 生理盐水，充分混匀，800 转/分钟离心 3 分钟，弃去上清，重复上述条件离心 2 次。最后弃去上清，加 GKN 液 4 mL，离心。

（3）弃去上清，加 GKN 液，制成 10% 细胞悬液。

（4）取上述细胞悬液以血细胞计数器计数，用 GKN 液将其调整为 1×10^6 个/mL。

（5）取以上细胞悬液 1 mL 于离心管，放入 37 ℃水浴中预热。同时将 50% PEG 液预热 20 分钟。

（6）20 分钟后，将 0. 5 mL 50% PEG 溶液逐滴沿离心管壁加入 1 mL 细胞悬液中，边加边摇匀，然后放入 37 ℃水浴中保温 20 分钟。

（7）20 分钟后，加入 GKN 溶液至 8 mL，静置于水浴中 20 分钟左右。

（8）800 转/分钟离心 3 分钟，弃去上清，加 GKN 溶液再离心 1 次。

（9）弃去上清，加入 GKN 液少许，混匀，取少量悬浮于载玻片上，加入 Janus Green 染液，用牙签混匀，3 分钟盖上盖玻片，观察细胞融合情况。

（10）计算融合率。融合率 =（视野内发生融合的细胞核总数/视野内所有细胞核总数）×100%。

【思考题】

进行异种细胞的融合有什么意义？

XXXI. Fusion of Animal Cells

【Objectives】

(1) Understand the basic principle of PEG induced cell fusion.

(2) Preliminarily master the cell fusion technology through the experiment of PEG induced chicken red blood cells fusion.

【Principle】

Under the action of inducers (such as Sendai virus, polyethylene glycol, etc.), the cells close to each other agglutinate, and then a series of changes of plasma membrane components occur at the plasma membrane contact, mainly the breaking and rearrangement of some chemical bonds, and then the cytoplasm mingle together to form a large dinuclear or multinuclear cell (called homokaryon or heterokaryon at this time).

【Apparatus, materials and reagents】

1. Apparatus

Microscope, centrifuge, balance, spoon, weighing paper, syringe and hemocytometer.

2. Materials

Venous blood of one-year-old rooster.

3. Reagents

(1) 0.85% normal saline, double distilled water and Janus Green dye.

(2) Alsever solution: 2.05 g glucose, 0.80 g sodium citrate, 0.42 g NaCl, dissolved in 100 mL double distilled water.

(3) GKN solution: 8.0 g NaCl, 0.4 g KCl, 1.77 g $Na_2HPO_4 \cdot 2H_2O$, 0.69 g $NaH_2PO_4 \cdot H_2O$, 2 g glucose, 0.01 g phenol red, dissolved in 1 000 mL double distilled water.

(4) 50% PEG solution: weigh a certain amount of PEG (Mr = 4 000) into a beaker, heat it in boiling water bath to resolve it. When it is cooled to 50 ℃, add equal volume of GKN solution preheated to 50 ℃, mix them well, and set it at 37 ℃ for standby.

【Procedures】

(1) Take 2 mL of chicken blood from the inferior wing vein of rooster; add it into 8 mL Alsever solution to make the ratio of blood and Alsever solution reach 1 : 4. After mixing, it can be stored in the refrigerator for one week.

(2) Take 1 mL of the stored chicken blood. Add 4 mL of 0.85% normal saline into it, and mix them well. Centrifuge the solution at 800 rpm for 3 mins; discard the supernatant. Repeat the centrifugation twice under the above conditions. Finally, the supernatant is discarded, and add 4 mL GKN solution into it. Centrifuge the solution again.

(3) Discard the supernatant and add GKN solution to make 10% cell suspension.

(4) Count the above cell suspension with hemocytometer and adjust to 1×10^6 cells/mL with GKN solution.

(5) 1 mL of the above cell suspension is put into a centrifuge tube and preheat the tube in a 37 ℃ water bath. Meanwhile, 50% PEG solution is preheated for 20 mins.

(6) After 20 mins, 0.5 mL of 50% PEG solution is added to 1 mL cell suspension drop by drop along the wall of the centrifuge tube with shaking, and then put the tube into 37 ℃ water bath for 20 mins.

(7) After 20 mins, add GKN solution to 8 mL, and put it in water bath for about 20 mins.

(8) Centrifuge the tube with 800 rpm for 3 mins. The supernatant is discarded, and GKN solution is added. Centrifuge it again.

(9) Discard the supernatant, add a little GKN solution and mix them well. Take a small amount of solution to suspend on the slide. Add Janus Green dye, mix them well with toothpick, cover the slide for 3 mins, and observe the cell fusion.

(10) Calculate the fusion rate. The fusion rate = (the total number of fused nuclei in the visual field/ the total number of all nuclei in the visual field) ×100%.

【Question】

What's the significance of heterologous cell fusion?

参考文献

［1］魏群．分子生物学实验指导［M］．北京：高等教育出版社，1999.
［2］杨洪兵，潘延云．细胞生物学实验教程［M］．北京：高等教育出版社，2011.
［3］姜伟，曹云鹤．发酵工程实验教程［M］．北京：科学出版社，2014.
［4］阿恩特，米勒．现代蛋白质工程实验指南［M］．苏晓东，曾宗浩，杨娜，译．北京：科学出版社，2016.
［5］贾士儒．生物工程专业实验［M］．2 版．北京：中国轻工业出版社，2010.